BEYOND THE FARM GATE

BEYOND THE
FARM GATE

The Story of a Farm Boy
Who Helped Make the Saskatchewan
Wheat Pool a World-Class Business

E.K. (TED) TURNER

University of Regina Press

Printed and bound in Canada at Webcom. The text of this book is printed on 100% post-consumer recycled paper with earth-friendly vegetable-based inks.

COVER AND TEXT DESIGN: Duncan Campbell, University of Regina Press.
EDITOR FOR THE PRESS: David McLennan, University of Regina Press.
COPY EDITOR: Brian Mlazgar.
COVER PHOTO: Chris Leachman Photography/iStockphoto.

LIBRARY AND ARCHIVES CANADA CATALOGUING IN PUBLICATION
Cataloguing in Publication (CIP) data available at the Library and Archives Canada web site: *www.collectionscanada.gc.ca* and at *http://www.uofrpress.ca/publications/Beyond-the-Farm-Gate-The-Story-of-a-Farm-Boy-Who-Helped-Make-the-Wheat-Pool-a-WorldClass-Business*

10   9   8   7   6   5   4   3   2   1

University of Regina Press, University of Regina
Regina, Saskatchewan, Canada, S4S 0A2
TEL: (306) 585-4758 FAX: (306) 585-4699
U OF R PRESS   WEB: www.uofrpress.ca

The University of Regina Press acknowledges the support of the Creative Industries Growth and Sustainability program, made possible through funding provided to the Saskatchewan Arts Board by the Government of Saskatchewan through the Ministry of Parks, Culture, and Sport. We also acknowledge the financial support of the Government of Canada through the Canada Book Fund for our publishing activities. We acknowledge the support of the Canada Council for the Arts for our publishing program.

This book is dedicated to my wife, Patricia Melville Turner, Mel, who has been with me intensely for over 63 years, and whom I have known for over 70 years. Without her willingness to sacrifice, I would never have had so many opportunities for service. I rescued her from her interesting job as a bank teller, hours 9 a.m. to 3 p.m., and took her to our farm, where she had the privilege to work from dawn to dark, and even longer. She raised our three lovely, competent daughters Jan, Joy, and Jill while I was often away on business. She cared for me as no one else ever could, and created a home environment where love flourished and responsibility was understood and accepted. When she was able, she joined me on work-related travels. Her presence was a tremendous added support for me, and she was a great public relations asset for the Saskatchewan Wheat Pool.

Thank you, Mel! I am forever grateful.

CONTENTS

INTRODUCTION

The story you are about to read is remarkable for many reasons, not the least of which is the fact that it illustrates the intimate interrelationships between biography and history. As is always the case, the significance of a biography, in this case that of Edward Kerr (Ted) Turner, is enhanced by considering the historical period within which the story unfolds. Chapters 1, 2, and 3 of this book provide an account of the courageous decisions made by members of the Turner and Osler families to immigrate to Canada from England and Scotland with an eye to establishing family farms in what is now Saskatchewan. It is important to recall that these decisions were in part prompted by the "National Policy" enacted by the Government of Canada in the latter 1800s. This policy largely focussed on industrial development in Central Canada, on trans-continental railway construction, and on immigration to the West. Land in the West was made available to settlers in order to establish a population base engaged in agriculture, a base that would both produce cash crops for export and be consumers for Central Canada's tariff-protected industrial goods.

The hardships, hard work, dedication, challenges, and triumphs of the thousands of settlers are legendary, but they are given 'flesh and bones' in Chapter 2, "Tom and Jessie." The cooperation and community building that characterized prairie life in the pre-Depression era included the emergence of concerns among farmers with regard to the operation of the private grain trade. The subsequent rise of agrarian political action and an associated cooperative movement culminated in the establishment of the Wheat Pools. Ted notes his father's involvement: he was "[a] strong advocate for change in marketing practises ... a leader and canvasser for the Saskatchewan Wheat Pool, and a loyal member as long as he lived. He was chairman of the local provisional board and over the years a leader in the local Wheat Pool." This involvement would have a profound and lasting effect on Ted. Even as he overcame physical and health challenges, Ted began to understand the importance of working together for the common good from an early age. We meet Ted more fully in Chapter 3 and learn how his family's emphasis on education, the rewards of hard work, and the benefits of cooperation all played a role in his life, and how these elements together encouraged him to attend the University of Saskatchewan.

The story of Ted's engagement with the love of his life, Melville (Mel), and her contribution to the establishment of their farm is touching, heart-warming, and typical of the experience of second-generation settlers in the region. And, as stated above, everything in Ted's history pointed to involvement in the co-op movement, and the events recounted in Chapter 4 once again reflect the intersection of history and biography. Ted made a choice to follow his family's ethical commitment to community, fairness, and social justice, and this, plus a historically specific constellation of economic and political forces, placed him on a path for early involvement in the Saskatchewan Wheat Pool.

All serious students of democracy and political processes will want to pay particular attention to Chapters 4 and 5. The notion of a mass-based, complex organization, in this case the world's largest grain cooperative, operating in a manner that maximized member democratic input and control may be difficult for the 21st century mind to fathom, yet that was the dream of the founders of the Wheat Pools. Ted's account of learning the complexities of the organization and the need to reconcile and mediate multiple stakeholder interests provides a lesson for anyone attempting to build contemporary citizen-based democratic institutions.

One of the many tensions inherent in a co-op movement in a market society relates to maintaining the original vision of such an enterprise while operating in the context of orthodox capitalist market practises. Capitalism is widely understood as a dynamic system driven by a competitive ethic that essentially requires enterprises to either expand and grow or face stagnation and failure. Cooperative enterprises, including the Saskatchewan Wheat Pool, were not exempt from these market-related pressures as we learn in Chapter 6, which describes why the Pool and its members found it necessary to diversify and expand its range of activities, not always with the expected outcome. The question that inevitably surfaces is how does such a complex organization maintain its relationship with its farmer base, a base that itself was undergoing a radical transformation.

The history of the prairies is replete with grassroots organizations and institutions serving the various needs of farm families and their communities. There was a pattern of overlapping involvement as illustrated by Ted and Mel's multiple organizational memberships and involvements. Although initially founded as a marketing cooperative, it is not surprising that the Saskatchewan Wheat Pool would be heavily involved in debates concerning many aspects of public policy. Perhaps there has been no issue, Medicare aside, which has stimulated as much debate on the prairies over the last century as the Canadian Wheat Board. Chapter 7 explains how and why the Saskatchewan Wheat Pool and its president, Ted Turner, were front and centre in the debates and activities that sought to support and upgrade the CWB, and why they staunchly opposed efforts to eliminate or diminish it.

As was noted above, the arrival of Ted and Mel's families in Western Canada was more by design than accident; that is, it was more a result of incentives provided by the National Policy of the Canadian Government. And part of that policy was the establishment of a captive tariff-protected market for Central Canadian industries, a market largely composed of family farmers. Those same farmers were to pay for the products of Canadian industry by virtue of the sale of cash crops on the international market. As such, the concept of 'globalization'—although the term was not yet a part of everyday discourse—could be applied to aspects of the agricultural economy from the early days of Western settlement. As Chapter 8 explains, a keen interest in the global wheat and grain market was always a priority for the Wheat Pool as it sought to maximize returns to its members.

Although our biographies and history intersect, and although we make history even as we are made by history, many changes occur behind our backs, often without our full understanding. Even as Ted grew and developed in his self-confidence and stature as the president, and even as Saskatchewan Wheat Pool evolved and developed into an enterprise the complexity of which could not have been imagined by its founders, other changes were afoot. In 1941 there were nearly 139,000 farms in Saskatchewan, but by 1991 there were fewer than 62,000, and during that same period the average farm size increased from 432 to 1,091 acres. In addition to changes in size, the farm community was increasingly differentiated by commodity, capital investment, and the use of hired labour to name just a few of the factors at play. One could legitimately argue that in the pre-WWII period there was a significant degree of homogeneity among farmers that made policy debates less conflictual, while by the end of the century farmers and the farm community were increasingly heterogeneous and differentiated.

As one might expect, such differentiation resulted in policy differences within the community and less agreement when it came to making agricultural policy recommendations to governments. As Chapters 9, 10, and 11 illustrate, this was precisely the environment that Ted and the Wheat Pools faced. On some matters, securing agreement among members of a diverse community was quite easy. This was the case with the efforts of the prairie Pools to compete and secure maximum benefits for their members in the world of larger and larger corporate enterprises that were operating on a global scale. This is explained in Chapter 10. With other matters, however, the debate was more difficult and contentious. This was the case with issue of the Crow Rate (Chapter 11), which brought out differences among commodity producers and provided strong indications that the federal government was adopting a new position in relation to supporting the traditional family farm community. At a personal level, for Ted the fragmentation of the farm community along various fissures meant that there would be opposition to whatever direction the organization went and conflicting policy positions on virtually any issue. In Chapter 9 Ted explains the dynamics of his decision to step down noting that: "It was a traumatic experience to leave the Saskatchewan Wheat Pool some 29 years after first becoming a delegate."

Agriculture and public service were, it is safe to say, literally in Ted Turner's blood, and as one might expect his departure from the Saskatchewan Wheat Pool marked the opening of other doors. Chapter 12, "Ancillary Companies," records Ted's activities with a number of other companies and with the Conference Board of Canada.

The word 'citizen' hardly captures the spirit of Ted Turner, but it may be the best we have. Always a proud alumnus of the University of Saskatchewan, he played an active role in fundraising and was awarded an Honorary Doctorate in 1989. Further, as he notes in Chapter 13, he was Chancellor for two terms, from 1989-1995. Chapter 14 contains the details of many other community contributions and activities, as does Chapter 15, "Reflections." Bitter sweet might describe the pride and sense of accomplishment tinged with regret contained in these "Reflections"; however, surely the bitterness becomes sweet when we contextualize biography. As was noted, Ted's parents came to Canada at a time when the federal government was engaged in efforts to build a national economy. In *Lament for a Nation* George Grant argues that Canada's nation-building project waned after the Depression and WWII when federal governments of all stripe turned to continental integration, a move that made the Western family farm a less important market for Canadian industry. Among the forces at work behind the scenes was a slow, subtle but important shift in state policy as neo-liberal cost-cutting and more faith in the blind forces of the market meant less support for traditional family farming. There were economic and political forces at work that were undermining the very foundation of the family farm and the Pools that served that community. Although not central to this story, these economic and political changes were to eventually swamp the Saskatchewan Wheat Pool. In short, history can momentarily override biography, even if it was not clear to those actors fighting, in this case, to advance the interests of family farming.

It has been said that humans make history, but not with a free hand. The Turner family story illustrates this fact. This family was an active participant in one of the greatest human migration stories ever told, a story that saw the transformation of the Canadian prairies into a stronghold of family farming. This is a story that is not without victims and tragedies; nevertheless, it is also a story of accomplishment, community building, cooperation, caring for others, and hope for the future. It is also a story that identifies the limits, constraints, and possibilities that are imposed by the historical period in which real-life biographies and stories unfold. Even as Ted and those around him had a hand in the development of the Canadian economy and important national and global political policies, it was not a totally free hand. The Ted Turner and Turner family story is one of courage, accomplishment, challenge, caring for community and others, and all the other elements of any human story; however, this story also illustrates the pressures, opportunities, and constraints that history imposes on each of our biographies.

Murray Knuttila
Professor of Sociology
Brock University
St. Catharines, Ontario

CHAPTER 1
Family Origins

As Thomas left the post office in Menith Wood, England, that June day in 1902, he went through the market square and gravitated to the public bulletin board, where he once more read the poster urging people to join a group of emigrants under the direction of the Reverend Isaac M. Barr. The group would travel to Canada in the spring of 1903, and had been assured that land in the western region of that country would be available at very little cost. Thomas had read the sign many times and knew every word on it. It had a magical pull for him, and an inward resolve was growing to take on this adventure. There seemed to be little future for him in Menith Wood where he lived with his parents, Arthur and Ellen Turner, three sisters and a brother. Another brother, Phillip, had left four years earlier, following the siren call of gold in the Klondike, in Canada's Yukon Territory.

The two oldest daughters, Florence and Mabel, worked in the village, while the youngest children, Molly and Bert, were still in school. Upon reaching the age of 16, Thomas was given the task of collecting the mail on Tuesdays and Fridays in Menith Wood. He would walk the three miles to the village and usually took time to go through the village square. It was on such an occasion that he first saw the settlement poster.

1

Turner family, circa 1890. Standing, left and right: Mabel and Florence.
Seated, left to right: Thomas, Ellen, Arthur, and Herbert.

In 1899, the Turner family was forced to leave its farm in Hereford-shire, just outside of Menith Wood. Arthur Turner was a charming man, more adept at social graces than agrarian pursuits, and things began to go badly when the farm manager, John Watson, was killed by a prize Hereford bull. Watson was very fond of 17-year-old Thomas, and was teaching him the skills of animal care and farming. When Watson died, Thomas pleaded with his father to allow him to run the operation but Arthur felt that his son was too young to look after the farm, and instead appointed Daniel Scott as manager. Scott made a real mess of things, and the farm was soon beyond solvency. Arthur and Ellen managed to salvage just enough from the wreckage to purchase "The Fox," a nearby abandoned pub, which they were able to convert into a modest home.

With the loss of the family farm, so, too, did Thomas lose his hope of becoming an independent farmer in the region. He felt he was doomed to live either as a farm labourer or some form of manual worker, but he wanted more out of life. Although he had been an excellent student, he loved the land, and he had declined the chance to obtain a post-secondary education. Having moved into the village, his twice-a-week task of collecting mail was made much easier, as he now had to walk a mere two hundred yards, rather than three miles.

Thomas lived in constant hope that one day he would carry home a letter from his elder brother, Phillip, whose last letter had arrived in 1899. In it, Phillip reported that he had filed a new claim and was sure he would strike it rich. Ellen wrote to Phillip every month for the next three years, but there was no response—not even a *"not known here."* Sadly for the family, and especially for his mother, Phillip's fate was unknown, and he was never heard from again.

Joining the "Barr Colonists," as the emigration group was known, would cost some £300—£400 to cover the cost of travel, meals and getting started, and an additional £100 to cover other, unforeseen, expenses. This was a staggering amount—over £41,000 in 2013 currency. Thomas's parents were unable to help, and even if they had the means, his mother would never have agreed to provide assistance, feeling that she had already lost one son to Canada. If he wanted to realize his dream of emigrating to Canada, Thomas would be forced to find the means to do so by himself.

Fortunately, a nearby farmer, John Saunders, who lived some four miles out of the village, was determined to upgrade his farm by draining a bog area and bringing another 50 acres under cultivation. To do so required digging a trench some 500 yards long, and up to six feet deep in places. Thomas signed on as a worker and walked the four miles, there and back, each day to earn the four shillings per hour. Most days, he spent 10 hours moving the heavy soil, and then on weekends, he was able to hire out to a nearby cricket team for £10 for Saturday and Sunday. He was regarded as the best cricketer in a wide area around Menith Wood, and, some claimed, in the whole county. He calculated that by March 1903, he would earn approximately £450: £10 per week digging the ditch, for 20 weeks, would bring in £200, while earning £10 per weekend playing cricket, for 25 weekends, would yield another £250. This would seem to be enough to enrol in the Barr Colonists. Luckily, he had saved some of his cricket money from previous years, so he was able to make the down payment of £50 by September 1, 1902.

Then it was time to tell his family. It wasn't easy, and everyone reacted as expected. Arthur said little and simply asked if he had thought it through and was prepared to accept the many challenges and pending hardships. He was inwardly very proud of Thomas, and recognized that his son was undertaking a venture that he himself had lacked the courage to do.

Ellen was instantly moved to tears, and tried to forbid him to go. Eventually, however, she recognized that Thomas's determination was too strong to deter; as well, she was inherently thrifty, and realized that if he didn't go, the £50 deposit would be forfeited. Florence and Mabel tried to calmly talk Tommy into changing his mind, while Bert swallowed hard and said, "Good luck, Tom." Molly looked startled and blurted out, "It will be great to be rid of a pesky brother!" She then dashed from the house and spent the next two hours sobbing uncontrollably in the woods on the edge of town.

The days rushed by and the money accumulated. Farmer Saunders offered Tom a good position on his farm if he would stay, and the cricket team was willing to double his weekly earnings. Ellen and Arthur contributed sound advice about a world unknown to Thomas. And Molly would scarcely let Tommy out of her sight when he was at home, but his mind was made up. Before embarking in early March 1903, Thomas had promised to write often, and to return in five years if he could do so. He promised Molly and Bert, who had become intrigued with his undertaking, that he would send for them when he became established in the New World.

THE ADVENTURE

The year sped by, and it was soon time to leave for Liverpool, where the SS *Lake Manitoba* would be waiting to take Thomas and the 2,000 other Barr Colonists to New World. It was a tearful goodbye as the whole family and many friends gathered on the station platform for the final farewell. Little did the family realize that, except for Molly and Bert, they would never see Thomas again. There were generous baggage allowances, so Tom could take with him all that was offered. In spite of her early opposition to the venture, Ellen became the consummate mother and insisted that Thomas take everything that he could carry, and would have bought new items if he hadn't objected.

The departure time was delayed some six hours to allow all the luggage and household effects to be safely stowed on board, but eventually the *Lake Manitoba* set to sea, and Thomas caught what would be his last sight of England. Once under way, he was constantly exploring his new

The SS *Lake Manitoba* in March 1903.
Courtesy of the Saskatchewan Archives Board, R-A12035.

surroundings. He had never been on a ship before, let alone have it as his home for the next two weeks. Like most of the other passengers, he succumbed to the constant motion of the vessel and was seasick for almost three days early in the voyage. Once recovered, he felt fine, except for a feeling that tugged at him constantly and that would remain for many months. He was startled when he realized that this discomfort was home-sickness. In spite of the trauma of leaving home, however, he really enjoyed the trip and actually made a few shillings in wrestling matches.

It was a thrill to see the approach to Saint John, New Brunswick. He had expected Montreal, but docking orders were changed. The next two weeks were a blur—finding all his belongings on the dock, going through customs, and delivering his property to the point at which it would be would be forwarded to Saskatoon. A special train, the fourth of the expedition, was required to move all the colonists' belongings. Thomas found a spot on the second car of the second train and before long he was underway, travelling across the vast stretch of land that was Canada.

In his passenger car Tom was befriended by an emigrant family—a mother and father in their early 40s and three teenage children. They treated Tom as one of their own and shared tea that was made on one of the two small stoves in the car. Tom felt comfortable with them, and it eased somewhat the departure from his own family. It was also reassuring that they seemed to know what to expect when they arrived in Saskatoon. Nothing, however, could have prepared them for the news that awaited them at their destination.

Barr colonists, camped at Saskatoon, 1903.
Courtesy of the RCMP Heritage Centre, 1997.47.3.

The Reverend Isaac Barr, to the consternation of the colonists, had absconded with the money he held in trust for them. Pandemonium erupted. The colonists searched in vain for those responsible, and simultaneously had to find food and shelter for their families—all of which Barr had promised would be awaiting them when the train arrived at its destination. Still a small community, Saskatoon could not properly house the colonists, and most were forced to purchase canvas tents. They also had to find provisions to tide them over as they established their farms: horses, oxen, carts, household supplies, food, clothing, and kerosene. Like most of the others, Thomas had only a small amount of money on his person—£150 that he had worked so hard to earn had disappeared with the Reverend Barr, who died in his bed some 35 years later in Australia, at the age of 90.

Unencumbered by family, Tom had more options for recouping his losses than most of his fellow emigrants. He soon learned that work was available in Prince Albert in the lumber industry, and transportation would be provided to anyone who would sign on in the next two days. Thomas made a quick decision to become a lumberjack. The train carrying his belongings arrived the next day, so he said goodbye to his newfound friends, purchased enough food for a week, and left for Prince Albert with some other recruits. So, instead of ending up in what would become Lloydminster some 200 miles northwest, he went to Prince Albert, some 130 miles north of Saskatoon, arriving there in early May 1903.

Fortunately, Thomas was young, strong and resilient, and he was able to adapt quickly to a new lifestyle. He soon learned lumberjack skills and in a few weeks was quite adept with an axe, could handle horses, and over the next year, would master riding logs down the river equipped with a cant hook to prevent log jams. He avoided working in the saw mill, preferring the bush and river work. Like most newcomers, he was the butt of practical jokes pulled by the seasoned veterans, but he could laugh at himself, and was soon accepted as part of the rough-edged crew. As was the case in most work camps, fights among the men were frequent. While Tom never initiated a fight, nor did he back down when confronted. His fellow workers soon learned that although he was only 21, he could more than handle himself if he had to.

The bush would be Tom's home for almost two years. The pay was good and there was a strong demand for lumber, so there was no down time. Meals were provided and the fellowship was great. However, Tom held on to his dream of owning and operating a farm, so in January of 1905, he visited the Land Titles Office in Prince Albert and filed on a homestead in the Maymont district. In March of that year, he decided to walk the 150 miles to claim his land. While still some 10 miles from his destination, he became snow-blind from the early spring sun shining on a completely snow-covered land; fortunately, he came upon a farmstead. He was welcomed in and stayed until his eyes recovered. George Holler, the owner, and Tom became lifelong friends. A few days later, when he could see again, he completed his journey and found his land. Maymont would be his home until he died 58 years later.

Like his mother, Thomas was frugal and he had been able to save almost everything he earned, so as the snow melted, he purchased horses, a plow, a wagon, tools, and a small tent. The farm was on the northern valley side of the North Saskatchewan River, with a wonderful view. But views do not make a living. Filing on a homestead, especially at a distance, is something of a crap shoot. This was no different. The land lay nicely with good drainage, and the soil was good, but it was infested with rocks, so every year Tom was obliged to remove the rocks that the frost had brought to the surface, only to go through the same, back-breaking process the following spring.

Between breaking the minimum 10 acres required of all new homesteaders—a task made more difficult because of the rocks—building a shelter for himself and his horses, and attempting a garden, the days were filled to overflowing. But by turning that first furrow some 100 yards in length, and seeing the rich black earth, knowing it was his, and that his success was in his own hands, he was hooked forever. Never would Thomas do anything other than farm, in spite of frost, drought, poor prices, grasshoppers, winter storms, and isolation.

The spring, summer, and autumn months were very busy and provided little time for anything other than farm-related activities. Fortunately, these included periodic visits to Maymont, some five miles away. Tom soon knew everyone in the village and made lifelong friends. This social interaction increased significantly in the winter months, and overnight visits to neighbours were a common occurrence. He loved where he was and he loved what he was doing. In spite of the harsh climate and the lack of easy access to equipment and repairs, the new farm was productive and Tom's willingness and ability to work long hours resulted in "proving-up" his homestead in only a few years. He was able to acquire an adjoining quarter-section from a neighbour who was unable to master the art of farming. This allowed him to write a letter to Menith Wood, indicating things were in place to receive Molly and Bert. He filed on a nearby quarter section in Bert's name and eagerly counted the days until their arrival.

In 1907, Molly and Bert arrived. There was enough equipment to farm the three quarters of land and with Molly cooking, housekeeping, gardening, and milking, all of the brothers' energy went into farming activities. Beef cattle were added—there was lots of open range—and timely farm operations resulted in excellent production. One of the big challenges each year was to have the crop threshed, and often the threshing was delayed until it suited the owner of the threshing outfit.

Tom and Bert soon made a major decision and purchased a huge Case steam engine, capable of pulling a 10-furrow plow, and powering a 48-inch Case threshing machine. Turner Brothers were now in the business of breaking land and of threshing crops; each activity generated good earnings.

THE SCOTTISH CONNECTION
Meanwhile, a Scottish family in Edinburgh was holding a pivotal discussion. Alex Osler, a draftsman for an architectural firm, had been suffering from poor health. His most recent visit to the doctor had resulted in the diagnosis that to overcome his respiratory problems, Alex and

Turner Brothers threshing outfit, circa 1910.

his family should move to either Australia or Canada, as the cool, damp climate of Scotland would only aggravate his condition. His wife, Cecilia, and the four children—Jessie, Robert (Bob), Cecilia (Sissy), and Alex—listened attentively as Alex Sr. explained the situation, and the conclusion was that they would move. Jessie was reluctant to leave her job and her boyfriend, Jock, but family was more important. Bob was doing preliminary schooling for dentistry but was willing to turn his back on a potentially promising career. Sissy and Alex didn't want to leave their school friends, but were intrigued by the "new" adventure.

The discussion rapidly moved away from, "shall we move?" to "where and when?" Australia offered the drier climate and had seaside beaches which were a family favourite, but it was so far away and the ocean voyage was daunting. The Canadian prairies had the desired climate and were much closer to the bonnie homeland. Canada was chosen and plans were started for the earliest possible departure in the spring of 1908. The land settlement office was contacted and available land was located near Spinney Hill, Saskatchewan. The farm was very isolated, in a heavily-treed area, difficult for cereal production but ideal for raising cattle. Alex Sr. would never get the hang of farming, but Bob seemed a natural, and over the years he developed an outstanding shorthorn herd and was very successful. His sense of humour served him and his neighbours well, and his close attention to public affairs made him an opinion-setter in the district. Sissy went to North Battleford and soon married Gordon MacAdams. Alex Jr. worked around the district and later found employment and comfort in the more northerly reaches of Canada.

Jessie, in the meantime, helped at home and taught piano to children in the district. She sorely missed the social life and business of Edinburgh and her many friends, especially Jock, of whom she was very fond. However, she knew that phase of her life would never be revisited, and she accepted her responsibility as the eldest child, to see to her parents' welfare. She was eager to attend every social event in the district, and at the Valentine's Day Dance in 1909, she met Tom Turner and was quite taken by his good looks and humorous personality.

However, the mighty North Saskatchewan River separated their respective homes and made contact difficult. Tom was also quite taken by the lovely Jessie, and in spite of the river, a romance began, culminating in their wedding on March 23, 1911, the date carefully chosen so travel across the river was at its easiest. They would be blessed with 50 years less two weeks of married life.

CHAPTER 2
Tom and Jessie

FARM LIFE

In 1909, Tom turned over his homestead and adjoining farm land to his brother Bert and purchased a half-section of superior land, S ½, 17-41-12-W3, some seven miles away, where he built a house from lumber, albeit quite small. In 1910, he rented an adjoining half-section and the following year, signed an agreement to purchase the N ½, 17-41-12-W3. He would later acquire the NE, 7-41, 12-W3 from the Canadian Pacific Railway (CPR), and later still, buy SE, 8-41-12-W3. This gave him 650 acres of cultivated land as well as pasture, with a spring and hay land. It was a very productive unit and one of the best farms in the district. The farmstead was about three and a half miles from Maymont, which became a thriving village and was soon larger than the surrounding communities.

Tom and Jessie worked hard and developed an excellent herd of dual-purpose shorthorn cattle. They raised pigs, ducks, turkeys, and chickens, had a huge garden, and their good farming practices yielded excellent grain and hay production. They also worked hard to improve their property. They planted a Manitoba maple shelterbelt with generous room for gardens and a grassy yard. The shelterbelt also served to separate the house from the barnyard. A dam was built that would provide ample water for the animals. The reservoir was eventually 15 feet deep and extended back into the farm a quarter of a mile. Spring run-off provided the water to fill it and its size ensured a year-round supply. For many years, it was also used by the community as a swimming facility and skating rink. A barn was built out of logs that came down the river some two miles way. One spring, Tom swam out into the river, attached a rope to a raft of logs, and pulled it ashore with a team of horses. A similar attempt was made to land a raft of lumber, but the current proved too strong and the rope had to be cut before the horses were pulled into the river. The captured logs were up to 16 inches in diameter, and 40 feet long. The barn served its purpose for some 30 years before a new one, this time from reclaimed lumber, was built in 1942.

The young couple welcomed their first child, Thomas Alexander, on December 30, 1911. He was followed by Winnifred Ellen in 1915, Evelyn Mary in 1920, Joyce Cecilia in 1922, and Edward Kerr in 1927. The growing family quickly filled the small shack, and a lean-to was added to provide more room. Then, in 1922, a large, five-bedroom, one and three-quarter storey house was constructed. The bedrooms and a bathroom occupied the upper level, while a large kitchen and pantry, a dining room, and a substantial living room were on the main floor. Unlike many farm homes of that era, the Turner house had a full basement equipped with a furnace that provided central heating adequate for the coldest of days. Remarkably, since most Saskatchewan farm homes were not "electrified" until the 1950s, the house also had a 32-volt generating plant and was completely wired. The outer walls were four-inch lap siding and inside, the finish was lath and plaster. The construction was so good that 70 years later, there were hardly any cracks in the plaster. The house was so configured that there was a large summer room, or "veranda," within the perimeter foundation. It was enclosed by seven windows and a door and was a feature room of the house during the seven warmest months of the year.

It was a proud family that carried its belongings across the yard to its new home. The most precious article to be moved was five-month-old Joyce, in her basket. The Turner home soon became a gathering place for neighbours and friends. There was fascination with the electric lights and also with the large console radio. People came from miles around to listen to "Hockey Night in Canada" on Saturdays during the hockey season. I was the family's youngest child, and I was born in the house on April 6, 1927.

Tom and Jessie's house, 1926. The home was built in 1922.

* * *

As always, the passage of time brought changes. Tom's brother, Bert, decided that farming was not for him, so he sold his land, and with his wife Nelly and family moved to Vancouver. Molly married a man named Mike Byrnes and they lived on a picturesque—but not highly productive—farm half a mile from the ferry river crossing. Their farm was a stopping point for travellers from south of the river going to Maymont for supplies and services. Over the years, they became known widely in the district as Uncle Mike and Aunt Molly because of their kindness and hospitality. They moved to Maymont, where they spent the final days of their lives.

With Bert's departure, the Turner Brothers threshing outfit was reorganized and it became known as the "Big Four," jointly owned by Tom Turner, who was the manager, along with Andrew Reid, Andrew Melrose, and Billy Melrose. It did all the threshing for the owners and all nearby farms until the mid-1930s, when drought and gas power took over. The outfit was large and required as many as 25 people and 40 or more horses to operate. It had a cookhouse and a bunkhouse—it was quite an operation. Tom's records of operations were meticulous and the "outfit" yielded a good return to the owners.

While not necessary it was desirable to have a qualified steam engineer, so to that end Tom hired Ed Buhr in 1917. Ed eventually established a garage in Maymont—an important business in the town—and became a virtual member of the Turner family.

Tom and Jessie, circa 1918.

The farm continued to develop nicely under Tom and Jessie. The farmstead grew with the addition of bins and barns and a large, productive garden area. While the farm prospered, there was still time for community work. Tom was councillor and reeve of the municipality, and a promoter and exhibitor at the local agricultural fair. He served on the committee to build a curling rink and on the founding board of the community hall. He was always a strong supporter of education and extension programs. A strong advocate for change in marketing practises, Tom was a leader and canvasser for the Saskatchewan Wheat Pool, and a loyal member as long as he lived. He was chairman of the local provisional board and over the years a leader in the local Wheat Pool.

The Turner family, 1933. Back row, from left to right: Joyce, Tommy, Winnifred, and Evelyn. Front row, left to right: Jessie, Teddy, and Tom.

Jessie, while busy with five children and often two hired men, squeezed in time for the Homemakers Club and always attended the annual convention. She was curious and pursued every opportunity to learn. She was known far and wide as an excellent cook and her hospitality was unmatched. In later years, after the family responsibilities diminished, she was involved in virtually every local organization and was respected for her willingness to work, for her good judgement, and for her organizational skills.

AFTER THE CRASH

The stock market crash of 1929 and its far-reaching effects brought farm product prices to disastrous levels, and people were forced to survive on very little. Simultaneously, the 1930s brought year after year of drought and grasshoppers, and threatened the extinction of farm operations across the prairies. The whole industry was kicked into survival mode as farms and entire communities disappeared. However, the Turner farm, because of skilful practices, a sound livestock operation, a fair-sized body of water, and good soil managed better than most. A huge garden watered from the dam brought some revenue but mostly served to feed the family and needy neighbours. The batteries for the electric plant wore out and there was no money to replace them, so it was back to lamps and lanterns.

Each year there was enough wheat for the following year's seed, and to buy a few items. Also, it was a custom, once the snow was on the ground, that two loads, each of 60 bushels of wheat, would be taken to the mill at Radisson and ground into flour. The teams would leave about 5 a.m. and would return at 8 or 9 p.m. The next morning, the 100-pound bags of flour would be carried into the house and stored for use during the next year. Some flour would be left on one of the sleighs and during the next few days, it was delivered to neighbours who were less fortunate.

Even though money was scarce, Tom and Jessie were able to provide for the family and four of the five children graduated from high school. Trees were cut and used to fuel the furnace and the kitchen cook stove. No event went uncelebrated—birthdays, Christmas, New Year, Valentine's Day, Easter—all were cause for celebration and usually with outside guests. The piano was also a centrepiece of attraction for family and friends.

It was amazing that a young man without any fully developed skills could be so successful. Tom had arrived in Canada with virtually no money. His only resources were a strong body, an intelligent mind, and a willingness to work hard. His partner Jessie was very similar, and together they built a good farm and provided for and educated five children. They earned the respect of their neighbours and together they served their community. They were completely satisfied in their last years, not wealthy but comfortable; their children were doing what they wanted to do, and they enjoyed their grandchildren as they grew.

THE DIRTY THIRTIES

The drought of the 1930s caused great adjustment for everyone, and dreams of growth and expansion gave way to the realities of survival. Farm size was put on hold so that every effort could go toward meeting the basic needs of life and "making do" with what one had. Debt was to be avoided or at least minimized. But out of this harsh situation came a tempering of character, a discovery of resiliency and a resolve to overcome the adversity of the moment. Those who survived were stronger and more confident individuals, but perhaps a little more conservative in their activities. The 1930s were a crucible that produced the true Saskatchewan "never give up" spirit.

This was the background for the Turner children who, by the end of that decade, were starting to leave the warm, loving nest, each equipped with resiliency and a sense of independence but also with a strong feeling of compassion and cooperation. It was not unusual during that desperate period for men without means to arrive at the Turner farm seeking work or something to eat. Invariably, there was a large jug of milk, a loaf of sandwiches and occasionally a piece or two of pie provided in a cool area under the shade of a large cottonwood tree. I was often given the task of carrying the food to the men resting in the shade, but was always told to come straight back to the house. One day, I learned a lesson that stuck with me all of my life. At the supper meal, I announced that there had been four bums there that day. A very stern lecture followed, chastising me for using the term "bum." I was told that they were people as good as us, but simply down on their luck. Never again was that word used in a derogatory way.

The 1930s took a dreadful toll on many Saskatchewan families, both rural and urban, who struggled to meet the barest needs of survival.

But as necessity is the mother of invention, need is also the instigator of innovation. The "Turner Farm" changed, and in order to meet basic needs, many household comforts were set aside. The electrical generator and storage batteries were idled, and kerosene and gas lamps took their place. The garden was enlarged beyond the size of family requirements, and the excess production was sold or bartered. The dairy operation was expanded so that 10 to 12 cows were milked daily, with cream cheques helping to defray living costs, while the skim milk by-product served to feed a growing hog enterprise, thus generating more cash.

One federal government program allowed for a $5 per month supplement if one hired a person through the winter months. This gave rise to the only incident of deception that I can ever recall in which my father was involved. He hired Jimmy, the son of Jim Steele, and Jim Steele hired my brother Tommy. The resulting $5 per month was the total income of those two young men. One year, Tommy went to Prairie River and worked in a lumber camp. Times were desperate and money was in very short supply.

THE TURNER SIBLINGS

It was during the 1930s that our family started to grow apart. Tommy continued to be the main farm operator. He had a knack for mechanics and ran the tractor, but he was also an excellent horseman. Winnifred completed high school, went to Saskatoon and took a hair-dressing course, and continued to work at the Labelle Beauty Salon. Ev and Joyce completed their high school education in Maymont, Ev in 1938 and Joyce in 1940. Ev spent three years in nursing training at City Hospital in Saskatoon, and Joyce went to Brandon in 1941 to work and stay with the McLeish family. The next year, she returned, and took a hair-dressing course in Saskatoon.

In 1940, Tommy enlisted in the army, and returned from overseas in 1945. The next year, we operated the farm in a three-way partnership with our father. The 1946 crop was only average, and in 1947 there was virtually no crop, and so the farm was not an attractive place. As well, the war experience had taken a toll on Tommy, and he found an easy alliance with alcohol. This made it difficult to operate in a partnership. Tommy was not happy as a farmer and he realized that he could not devote his life to that occupation. Always happy, he had many friends, but many of them were also fond of the bottle. I was prepared to move on to another life—as a veterinarian, perhaps—but Tommy made the decision to join Evelyn, who married Bill Mitchell in 1948, on the ranch near Lloydminster. Isolated as it was, Tommy loved the life, riding every day, looking after cattle and worshipping his three nephews—Alistair, Jimmy, and Ian. Bill and Tommy got along great, even though they argued Army versus Air Force at great length. Eventually, the harsh Saskatchewan winters caught up with Tommy, so he would spend the colder months at Parksville, British Columbia. He died in January 1991, in his 80th year, of cancer of the lungs—the result of 65 years of cigarette smoke. Tommy was a brother whom I had worshipped as a boy, but because of our age difference of 15 years, and his five years away in the army, I was not very close to him. I respected him and we could have been good friends, but we had little in common. Tommy never married.

Winnifred married Bill Wilson of Saskatoon at the farm in 1938. Bill was an insurance salesman with Trotter and Company, and shortly after they married, he accepted a job with Stan Elliot in Rosetown. Wyn got a job in a beauty parlour in Rosetown and soon thereafter purchased it; it became "Wyn's Beauty Parlour." She operated it successfully for many years, eventually selling it to become involved in the family business that Bill had purchased from Elliot. Wilson Agencies operated for many years under the ownership of their only son, Tom.

Wyn and Bill also had a daughter, Dawn Elaine, who became a registered nurse, graduating from University Hospital in Saskatoon. Dawn was the flower girl for Mel and me and spent much of each summer at our farm. Dawn died at the age of 54, a victim of tobacco. She just couldn't shake the habit, even thought she well knew the consequences of smoking. Wyn developed severe allergies and suffered greatly in her final years, dying in May 1990 at the age of 75. Wyn had spoiled me as a child and was always very supportive of everything I did. We were as close as our age difference and distance allowed us to be. She was thrilled when I received my Honorary Doctorate and Co-operative Order of Merit, and was enthralled, along with the rest of my siblings, to witness my installation as Chancellor of the University of Saskatchewan.

Evelyn, upon graduating from City Hospital, nursed there for a while and then answered an ad for nurses from the Lloydminster Union Hospital. She was there when I was attending university and her occasional letters, with money, were a great boon to me. One of her patients was Bill Mitchell, and in April of 1948 they were married in Sharon

United Church at Maymont. Ev left nursing and, with Bill, they lived on a farm/ranch at McLoughlin, Alberta, and then took over the Battle River Ranch at Unwin. They had a great life with many close friends and livestock events, and in later years attended many air force reunions.

Their three boys kept them busy. Alistair established his own operation within the boundaries of the Battle River Ranch (approximately 7 miles x 3 miles). Jim built a house nearby, from which he operated a trucking business. Ian moved a trailer onto the main farm yard and lived there until he married, and then changed residences with Bill and Ev. Ian now owns and operates that portion of the original ranch. Erin, Alistair's daughter, manages the other portion.

Bill and Ev died within four days of each other in March 1997. (Bill's parents had died four hours apart in 1956.) Ev had been very athletic when she was younger. A good softball pitcher, she also excelled at track and field, and was active in curling and bowling in her later years. She won numerous ribbons at district school field days. She loved to dance and could play the piano, but only did so on occasion.

Joyce was a vivacious, auburn-haired girl with a hair-trigger temper in her childhood and young adolescent years. She was quite attractive and always had boyfriends. As a child of five years, she was somewhat resentful of me as the new baby, who had arrived and was taking attention away from her. However, before long, she was a loving, even devoted sister to the young invader of her territory. Joyce was active in track and field but not as athletic as Ev. Upon completion of her hairdressing course, she went to Rosetown to work for Winnifred. In 1945, she married Bill Evans, a farmer some seven miles north of Rosetown. They had a wonderful marriage and a host of friends. Vickie, their oldest, became a registered nurse, married, and moved to Toronto. Tim was mentally challenged, and in his 20s, was a resident of the School for the Mentally Handicapped at Moose Jaw. Joyce and Billy worked hard to achieve a special facility at Rosetown. Joyce spent her last years in Saskatoon and she died of breast cancer in June of 1990.

CHAPTER 3
Ted's Story

BACKGROUND INFLUENCES

Following the First World War, things were very positive in farm production and our farm was thriving. However, the early 1920s were also witness to great adjustments in the fortune of farmers in the prairie region. Farmers were sickened by the way they had been exploited, and decided to collectively do something about it. The result was the formation of a Wheat Pool in each prairie province. My father, Tom, was the chairman of the Provisional Board of the soon-to-be new Pool at Maymont. He had persuasive skills in drawing adherents to the cause, and would host meetings at the farm. In addressing the cause, he would say that, "the farmer is a most peculiar animal—it can be skinned more than once." Or, "if you put a banker, an elevator man, and a railroad man in a barrel and rolled them down a hill, there would always be a son of a bitch on top." The farmers were successful in organizing the Saskatchewan Wheat Pool, and Saskatchewan was positively changed forever. Years later, Tom would witness my own progression in the organization.

My father's involvement in the Saskatchewan Wheat Pool continued as long as he farmed and his loyalty never wavered. Not one bushel of grain went to any other company. Throughout the 1920s, my parents continued to develop their farm and their chosen community, taking leadership roles in various organizations, The Homemakers were a special passion for Jessie, and she attended regional and provincial conferences.

Toward the end of the decade, however, disaster struck Saskatchewan in several forms. The stock market crash in 1929 triggered financial instability in markets, resulting in farm commodity prices tumbling to unprecedented lows. The situation was further aggravated by several years of severe drought, dust storms and grasshopper infestations—the "Dirty Thirties." Few people had money and many struggled to feed themselves and their families.

It has often been asked why there has been a socialistic trend in Saskatchewan. First of all, Saskatchewan was the hardest hit by the Depression and the drought. Governments responded with token handouts of food and some anaemic attempts at job creation, but those in the rural areas—both farms and small towns—felt abandoned by those in power. Then suddenly, there was a war to fight, and immediately there was money for weapons, food, and armed services personnel. Why were those resources not available when people had been in great distress?

Just as the line elevator companies were rejected for their exploitation of farmers, so, too, were the political parties punished because of their indifference to people's needs during the 1930s. People said there had to be a better way, and flocked behind a dynamic young Baptist minister named T. C. ("Tommy") Douglas, and once again, the face of Saskatchewan was permanently and positively changed.

MY EARLY YEARS

As noted in the previous chapter, I was born on the family farm in 1927, during a period of comparative prosperity in western Canada, but I remember virtually nothing of this early period. I am what can be called a "child of the Depression." At age two and a half years, I was playing with my good friend Jimmy McLeish, who was some 70 days younger. A disagreement arose and Jimmy, upon gaining possession of a small metal toy threshing machine, struck me with the object at the base of my skull. The resulting coma lasted for a few days, and the effects of the concussion would take 10 years to overcome. The effects of the blow expressed itself in convulsions, with great frequency and severity. Doctors in Saskatoon were at a loss to know what to do. A Dr. Brown was particularly sympathetic and tried desperately to find, if not a cure, at least some way to modify the convulsions. He referred my parents to a Dr. James in Chicago.

Letters were exchanged and Dr. James prescribed medicine by long distance that would eventually rectify the condition. I well remember a brown pill twice the size of an aspirin to be taken daily, a capsule also for daily consumption, and a smaller pill to be taken as needed. I was never sure how mother made the determination to administer it. I sensed the near panic that would arise in my parents as the pills dwindled and the new supply had not yet arrived.

It would have been easy for my parents to confine me to the house, where I could have been under constant observation, but they desperately wanted me to lead a normal life. I was allowed, even encouraged, to go out to play with the "Denomy kids" who lived just across the road from the Turner farmstead. We would play by the hours, often making up games using whatever was at hand. Hide-and-seek was the most popular because there were so many neat places to hide. Gordon, the oldest of the six Denomy children, was some nine months older than me and had learned what to do when I took an epileptic seizure or "fit," as it was commonly called. I always had a spoon in my pocket, and Gordon would take the spoon and hold down my tongue during a fit to prevent me from swallowing my tongue and choking.

In retrospect, I marvel at how Gordon, at such a young age, was willing and able to perform the task. I am also sure that, unknown to me, my mother was often watching from the window as we played. I soon learned to recognize the approaching seizure that often followed a time of excitement or exertion. First came a sense of vagueness, then objects started to lose their form. Even on the brightest of days, everything would go black and I would fall down. I soon knew enough to either sit or lie down. Then my body would convulse, and at this point, my tongue would go back in my mouth. Some time later, I would regain consciousness very slowly, things would brighten, and objects would again become discernable. I had a sense of complete weakness and would lie for some time, fully conscious, before being able to stand. But, before long, I would be back on my feet and the interrupted game would resume. In the recovery process, as the darkness would lift, I always had the most terrible taste in my mouth, and it was the darkness and mustard-brown colour and the terrible taste that I can recall to this day.

To conclude this part of my story, I started school at the normal age of six, and had a Mrs. Martin as my teacher. She was a good friend of my parents, and she understood my situation and was prepared for it. I had lots of fits in my first two or three years at school, but the teachers and other students handled the situation. The seizures decreased in frequency as I grew, and by age nine, they had almost ceased; probably only one or two occurred thereafter. It is amazing that the other students never teased or ridiculed me for taking "fits." If anything, they were very accommodating. I have always been grateful for that!

The first year of school was difficult, and I think it was only through the good graces of Mrs. Martin that I was promoted to Grade 2. She probably felt it would be a discouraging blow to hold me back. Her judgement was well-placed, because once I gained confidence, school work was not a great challenge. The Grade 4 class was very large and so several of the best students were promoted into Grade 5. My mother asked the teacher to leave me in Grade 4—a good decision, because I probably didn't need the extra challenge. The result was that I was the top student in my class for the rest of my public and high school life.

Sports had always been important in the Turner household, and all the siblings were adept at track and field events, winning numerous ribbons from district field days. We had swimming opportunities at the dam on the farm—the same place that had a skating rink in winter. Right across the road, in the Denomy yard, was a ball diamond, complete with back stop. So, while I progressed academically, I also developed in ball and hockey while at school. Dad and my brother Tommy made a small rink behind our house where I spent countless hours skating and shooting. A lamp in the window extended the hours of use. In the summer months, I played catch constantly, with my siblings, with my friends at school, and with Gordon and Cecil Denomy. I would become a very good ball player. I spent hours hitting stones with a broken fork handle, thus perfecting eye-hand coordination. Jim McLeish and I especially would pitch to each other for long periods. This paid off when, on our school team, we would alternate between pitching and first base. I don't think we ever lost a game.

Every summer, the month of July was spent at Meeting Lake, some 50 miles to the north, the Turners in "Everest" and the McLeish family next door in "Neverest." It was a great experience each year—with swimming, ball in the meadow, fishing, miniature golf, and horse shoes.

CHAPTER 3

Time went all too fast, and soon we would be back home to prepare for the harvest. The Turner/McLeish relationship was unique. The two families were very compatible, and while not related, each seemed to be family to the other. Gracie and Mac, and Tom and Jessie would take holidays together, often camping trips. Doreen, the youngest, called "Dickie," spent the entire summer at the farm, only going back to May-mont as school approached.

Doreen was only nine years old when her family moved to Brandon in 1941; consequently, we only saw her on infrequent occasions thereafter. In her late teens, she married Jack Millar from Winnipeg, and Jack continued his career with Lever Brothers in Thunder Bay. A few years later, he was moved to Saskatoon as sales manager for Saskatchewan. Their three young children fit nicely with ours and we visited frequently between Saskatoon and our farm. This wonderful relationship continued after our move to Regina and my friendship with Jack grew; he was a superb individual and an accomplished golfer. Jack died at age 44, far too young, and Doreen was left with two young daughters and an even younger son. We were delighted when, a few years later, Doreen married Cy Rouse, and they melded their two families into one, with two boys and five girls. We still associate closely with them and stay with them in Waskesiu whenever I attend the Senior Lobstick.

My friendship with Jim McLeish continued until he moved to Brandon in 1942. Two years later, in 1944, he lied about his age and joined the Navy. Jim married and had two children, but contracted polio in 1952 and never walked again. He defied the experts who said he would never leave the iron lung and spent many years in a wheelchair, having learned how to gulp air and how to talk once again. Jim only had movement from the neck up. During his life at Winnipeg's St. George Hospital, he became a leading salesman for Amway products.

Upon becoming a director of the Saskatchewan Wheat Pool, I would, on occasion, be in Winnipeg. At the end of the day on my first time there, I took a taxi to the hospital to visit Jim, whom I had not seen since 1942. During the taxi ride, I pondered what I would say—me with everything, and him in a narrow world. Upon arriving at the hospital, I told the driver to take me back to the hotel. An hour later, ashamed at my cowardice, I was back at the hospital, where we had a wonderful visit. Jim was up to date on current events and was so pleased that I came. Thereafter, my visits were frequent. I recall one time when I

24

arrived at his room; he was so excited he could hardly contain himself. His wheelchair had been motorized and he could operate it himself by way of a crown-like device around his head. For the first time in years, he could go places on his own. Jim was an inspiration to all who took time to know him. He did not allow a devastating condition to destroy his life.

* * *

The lake was a great place with lots of other kids from the surrounding communities. I was fascinated by the dating activities of my sisters and watched with great interest the excitement around the weekly dance in the pavilion. It was probably here that I first started to understand girl-boy relationships. I was a slow learner, and coupled with a shyness caused by my illness, it would be many years before I had a girlfriend.

As my strength and confidence grew, I became useful around the farm and soon had specified jobs—carrying wood for the kitchen stove, and getting the cows into the barn for evening milking, a task which was helped by a riding horse and a good dog. I was expected to drive the team of horses as Dad and Tommy loaded hay by hand or spread grasshopper poison.

Horseback riding was something Gordon and I did almost every day in summer, and our games of cowboys and Indians seemed all the more real because of live horses—Silver and Scout. I always had two or three slingshots and was remarkably accurate with them. Bows and arrows we made met with only limited success. The reduced frequency and severity of seizures was a great incentive to play hard and to venture out on my own.

At about age 10, I was handed a pail full of strychnine-laced oats each Saturday morning and sent out to poison gophers. Saturdays were also occupied by picking potato bugs which, if left unattended, would completely eat the tops of the potato crop. I recognize now how skilfully my father eased me into farm work—weeding the garden, cleaning the barns, milking cows, and feeding hogs. These chores helped me develop the physical skills of a grain/livestock farmer, but even though the work load increased, there was still time for hockey in winter and baseball in summer.

I was 12 years old in 1939, the last epileptic fit was behind me, and adolescence was just around the corner. This year also saw the start of the Second World War, an event that, even though half a world away, would, in fact, effect everyone in Canada. In the next few years, our community would be stripped of the 18- to 40-year-olds, which was the group that provided most of the physical labour on the farms. It became necessary to replace them by the young teens and a group of older men.

TEENAGE YEARS

Maymont, like every other district in Saskatchewan from 1941 to 1945, saw most of its able-bodied young men and many women join a branch of the Armed Forces. Many women took up jobs outside the community to support the war effort. My brother Tommy signed up in 1940, and the farm was left without the necessary manpower. One man my father hired remained until 1941, and then he also signed up when the government, needing more soldiers, raised the upper age limit. Fortunately, we had purchased a rubber-tired tractor in 1938, a McCormick Deering W30, one of the first tractors on rubber in the district. Because of this my father, with the help of teenagers whom he was able to hire, was able to carry on with the field operations.

"Bacon for Britain" was a public exhortation and so we went into hog production. This was very labour-intensive at a time when help was hard to find, but it was the patriotic thing to do. The 1941 crop was poor, and there was still some manpower available, so it was not a huge challenge to harvest the crop. The McLeish family moved to Brandon in the late summer of 1941, as "Mac" was promoted to manage the Imperial Bank branch there. Joyce went with them to help the family get settled.

On the other hand, Jim stayed with us and he and I, at age 14, stooked our entire crop of 400 acres. Stooking involved picking up the sheaves that had been produced by the binder and placing them into piles called stooks—generally eight sheaves to a stook—so the crop would ripen and be protected against rainfall. Stooking was very hard work that involved stooping and lifting. It was no mean feat for two 14-year-olds to follow two binders. Dad and a hired man ran the binders. Threshing followed, but Jim went to Brandon and I went back to school, as we were considered too young for that heavy labour job. However, 1942 was another matter. Jim returned from Brandon for the harvest and again we stooked the whole crop, which was much more demanding than the year before, because the crop was so good—at least twice the yield of the previous year.

Dad had taken to growing barley to support the hog enterprise. Barley had slippery straw and a skin-irritating dust, so it was not pleasant to handle. By 1942, virtually all of the able-bodied men were gone, so Jim and I, at age 15, with two 16-year-olds, Nelson and Eddie Nutbrown, plus two men who were in their late 40s formed the threshing crew. We threshed for 30 days on the Turner, Nutbrown, Sloan, and Clayton farms. Dad, who was 64 at the time, hauled most of the grain away from the threshing machine, an amazing achievement. A typical day would see us out of bed by 5:30 to feed, water, and harness the horses. We would then eat breakfast, hitch the horses, and leave for the field where the threshing machine started at 7 a.m. and continued to operate until 7 p.m. with a one-hour stop from noon to 1 p.m. It was very heavy work loading the stooks onto the hay racks, and then unloading them into the thresher. Normally, one would load and unload once every hour.

Our bodies developed from the heavy work and our attitudes to hard work were formed in a very positive way. We were proud that we could do a man's work at age 15. In spite of the long hours, once the thresher stopped at 7 p.m., we still had to return to the barn and care for our horses—feed and water them, unharness them, then curry and brush them, after which we would have supper and then go bed, generally by 9 p.m. That year, we threshed on several Sundays, so we were deprived of a day of rest. Even at that, not all the harvest was completed. An ice storm followed by freezing temperatures on November 5 closed out the season.

That year, I returned to school for my first day on November 6. The teacher understood and held special classes for us on subsequent Saturdays. This was good of the teachers, but not much appreciated by us, since it meant losing a Saturday of leisure time!

The following years were repetitive, but in 1943, a poorer crop occurred. We threshed our usual farms and Jimmy Reid was part of our crew because Norman Melrose could not find enough help to start up his outfit. Ed Buhr also came to help us, and then he, Jimmy Reid and I went and threshed for 12 days for Norman. My wages were $3 per day plus $1 for the team and rack, so I earned $48 in total.

What a surprise I got when I met Norman in Maymont one day, and he was settling his harvest debts. It was the usual practise that farmers paid their bills after they had marketed some of the crop. He said he was well-satisfied with how the threshing had gone, and insisted that I take $50 for my time in his outfit. Today, $2 seems insignificant, but in 1943, it wasn't, and it was not common to pay more than was required. But even better, he told people around Maymont that I was the best man he had ever had on his threshing crew, and he had been harvesting for 20 or 30 years and had worked with many men. What a lift for a 16-year-old to realize that I could do a man's work! A passage, for sure, into adulthood!

In 1944 we again had a huge crop on the Turner farm, and the same challenge of finding enough labour occurred, so again the harvest took precedence over the school year. A bumper crop was considered to be 40 bushels to the acre of wheat, and we had fields that yielded 50, so there was a lot of hard labour involved, but it very satisfying.

As I moved through my teens and high school, I worked very hard on the farm, year-round. The "Bacon for Britain" hog production meant most weekends involved preparing enough grain to last the hogs for one or two weeks. This was a real challenge in winter, when just starting the tractor to grind the grain involved heating water to fill the tractor so it would start.

BUILDING CONFIDENCE

The need to do a man's work at age 14 was a real confidence boost for me. As an adolescent I was small for my age, which compounded the shyness I felt from never knowing when a seizure might occur. The hard work enhanced my physical development, as did my involvement in sports of all kinds. I was an average to good hockey player and an outstanding baseball player. All of this, combined with the fact that I generally had the best average of anyone in my class, allowed me to become increasingly self-confident, except when it came to boy-girl relationships. It would be another two years before I got up the nerve to ask a girl for a date.

It was in the early 1940s my sister, Evelyn, undertook to teach me how to drive a car and how to dance. I can recall thinking, "I know I will be a good driver, but I doubt that I will ever learn to dance!" I was prophetic on each count. Not being musical inhibited my dancing ability, and that, in turn, increased my shyness around girls. Dancing was a main social activity, and dates were often "for a dance." Fortunately my best friend, Nelson Nutbrown, who lived only two miles away (and who I really got to know when he graduated from Calais School and came to

Maymont to complete high school) was very fond of girls, and he taught me by word and example how to associate with them. So, by the time I was 16, I had started dating different girls.

POST HIGH SCHOOL YEARS

Following my graduation from high school, I remained home for a year to help on the farm. I was restless, as all my close friends were elsewhere, and in spite of an active involvement in baseball and hockey, and an introduction to curling, I felt lonesome. My father encouraged me to enrol in the School of Agriculture and I did so in the autumn of 1946. This was a departure, because even though I loved the farm, I had not seen my future there. I had wanted to be a veterinarian, but finances were tough and so I opted for the School of Agriculture as an interim measure.

The school was a wonderful experience for someone who had never been very far from home. The first year, I lived in residence at the airport in an H-hut that had housed airmen in training during the war. It was simple but adequate accommodation, with ample but not highly-attractive food. The rent was cheap, but there was an added cost for transportation. Each day, a ticket provided a bus to 33rd Street and then a streetcar to university; it took about one hour, if connections were good. Then, the reverse to get home in the evening. To go downtown was also a city bus trip. Taxi drivers, who had dropped a fare at the airport, would often pick us up at the bus stop, not start the meter, and drop us on a back street near city centre. They would be severely chastised if they were caught doing so. We really appreciated their generosity to a group of moneyless students.

The second year I boarded in a house on 4th Avenue near 25th Street, at a cost of $40 per month that included breakfast, packed lunch, and supper. Bob Wright and I shared a small room and the bed. Luckily, we got along just fine. With two weeks left in our university stay, Bob had an argument with the landlady, so she kicked us out. Fortunately for us, Bill Small phoned his landlady, and she took us in. What a change! The meals were much better, we were encouraged to sit around the living room in our leisure time, and Mr. Beyers was constantly playing jokes on us. The only shortfall was that it was too far away to walk to classes as we had done from 4th Avenue.

Students like me were very lucky to share the classroom with more mature men, many of whom had been away to war. They brought a good deal of wisdom to the classes and the instructors had to be on the bit because they were not dealing with some naive young people. We learned a lot from our fellow students.

There were lots of chances for sporting activities and we had the second best intramural hockey team on campus, losing 2-1 in the final to the College of Commerce. The leadership provided by directors of the School of Agriculture—Bill Baker during year one and Art Stilborn in year two—was outstanding and inspirational and did much to shape my philosophy about life, and to develop my attitude toward community. I can still hear Baker admonishing us to go home and fulfil our obligation of leadership. He said we had accepted the privilege of attending the university that was there because of tax dollars; therefore, we had the responsibility to give back to society in general.

The technical knowledge I gained was extremely useful, but paled in comparison to the personal growth, the broadening of outlook, and the development of interpersonal relationship skills that I achieved.

THE 1940s

In high school, we had a junior baseball team and Bergie Bergman, the bank manager who was raised in the United States, coached us and taught us techniques and strategy. We lost very few games, playing mostly at Sports Days. I, along with Duffy McDonald, provided the pitching, until one year, on a cold May 24, I pitched three five-inning games. We won them all, but for the next two weeks I couldn't raise my arm high enough to comb my hair. A good arm was ruined, and I was never able to pitch baseball effectively again. So, the switch was made to shortstop and second base. I developed the ability to switch-hit long before it was popular to do so.

Maymont had a long history of having a very good men's baseball team, except for the period during the Second World War when all the men were away. During that time, local tournaments featured youth teams, generally 16 years of age and under. We had a very good team, so J. D. Blacklock, who ran the tournament, raised the age each year to keep our team intact. We lost very few games either at home or away. When the men started returning from the war senior ball took over, but we continued with our younger team. In 1946, the men's team started using some of us juniors, and within two years a few of us were regulars on the men's team. I was used first of all in the outfield but soon settled at shortstop because we had a strong second baseman in Ches Miller. The Linnel brothers, Jerry and Mickey from Keatley, joined us and gave us pitching and hitting. I was soon batting near the top of the line-up, a singles hitter, right-handed, and a power batter left-handed. Baseball became my favourite sport and would continue to be as long as we played. Baseball eventually gave way to softball, as the number of teams playing baseball declined. We switched, and if anything, had an even better win-loss percentage in softball. I played until, and even a little after, we left Maymont in 1966.

The 1940s were the crucible of my life. I was 12 years old at the start of the decade and 22 when it concluded. Physically, I started as a 98-pound weakling and exited the decade as a powerful man due to the hard physical work I had done. Educationally, in 1940 I was apprehensive and lacked confidence. By 1950, I was confident—but not cocky—with a university diploma in agriculture. In terms of sports Maymont had a team in a hockey league in North Battleford, and after I left to go to university, I joined our School of Agriculture hockey team. All of this activity helped to develop my skills and coordination, so all in all I was quite a good athlete by 1950, by which point I began to find curling to be intriguing.

Socially, I was very reticent in 1940, but exited the decade self-assured, in love, and engaged to be married. My personal development aided my social progress and as I passed into adulthood, I thought of myself as cheerful, friendly, compassionate, respectful, and receptive. I had learned, by example, from my parents, and by personal involvement, of the value and satisfaction that came from community and volunteerism. I had spent many hours helping to tear down an air force building in North Battleford and then helped in rebuilding it into a magnificent community hall that served Maymont for many years. The development of interpersonal relationship skills would serve me well for the rest of my life.

But what happened to the shy, young adolescent that allowed him to realize his potential as a community, provincial, and even a national leader in agriculture? The 1940s were the development decade for my life. Exiting from the poverty of the 1930s, the anxiety of the Second World War, followed by the exhilaration of the end of the war, then

living in a vibrant community, and enjoying improved economic conditions—all made me into the man I would become. By 1949, I was in
partnership with my father and taking on the management and most of
the labour responsibilities of the farm.

I made friends easily, even with girls, as long as there were no roman-
tic implications. When I enrolled at university, I wasn't dating regularly.
Sports still dominated my activities. Interestingly, our group hardly
consumed any alcohol. Ches and Shirley Miller and Derby and Binky
Reid, each about 10 years older, would entertain us in their homes and
thus provided an excellent environment for us young people.

MELVILLE

Quite unplanned, while I was on the summer break from university, an
incident during September 1947 changed my life forever. My mother
asked me, one Sunday afternoon, to drop off a write-up for the North
Battleford newspaper with Mrs. Stan Bright, who was the correspondent for the North Battleford *News-Optimist.* This was the home of
Melville Bright, who was one of "our group." She was very personable
and easily the most popular girl in our group. She had gone out with my
good friend Alex Spence for some time and was currently going steady
with Don Stynsky; they were becoming less and less involved with "the
group." When I arrived at the Bright house, Melville was the only one
home. I delivered the letter, and since we had always been friends, we
chatted for a while, and I asked her if she would like to go for a ride. I
was startled but pleased when she agreed. So, we spent the rest of the
afternoon together. After that, we started to date.

During the month leading up to my return to university, we dated
often and I became increasingly fonder of Melville. I was reluctant to
go back to Saskatoon because, by this time, I didn't want to lose her,
and I was going to be gone for five months, even though I would be
home for two weeks at Christmas. I left with the promise that we would
correspond, which we did. In those days, there wasn't the easy back-
and-forth travel between Maymont and Saskatoon that was prevalent
in later years. I looked forward eagerly to letters two or three times per
week from this wonderful girl. Upon my return in March 1948, we dated
steadily and for her birthday in March 1949, she accepted an engagement ring from me. We made plans to marry in July 1950.

We thought that 1949 would be a year of great fun. I would be busy on the farm, and Melville with her job at the Imperial Bank as a teller, but there would be lots of opportunities to be together. What a surprise when the manager told her one morning in April that she was being moved to Saskatoon and was to be there on the following Monday. There were no other jobs available in Maymont, and her income was important to our marriage plans. Things had been tough on the farm, with a complete crop failure in 1947 and an average crop in 1948. So, she accepted the move. I got in touch with Mrs. Beyers, where

Courting days. Ted and Melville, 1949.

I had lived the last three weeks while at university, and she agreed that Melville could stay there until she found other accommodation. The Beyers were very accommodating and very hospitable, so the problem of housing was temporarily solved. However, the Beyers were in the process of moving, and in the new house, there would not be enough room for Melville. She soon found a place on 4th Avenue, which was close enough to the bank so she could walk back and forth to work. She made some long-term friends from among the seven girls who lived at that location, and still sees or corresponds with a couple of them. As well, she made very good friends with her fellow workers.

Melville greatly enjoyed her time in Saskatoon, and it was a good experience for her. Our letter-writing continued, but she was able to catch the bus most weekends, leaving on Saturday at noon for Maymont and returning on Sunday evening. She would either stay with her sister Shirley or, most often, came to our farm, which allowed Melville to come to know my parents better than if she had actually remained in Maymont.

We set our sights on July 5, 1950, as our wedding day, and this would be a beacon to guide us in the things we would do for the next year. I was very busy farming and playing at least one ball tournament per week, so the time went quickly, even though I desperately missed Melville. Once the harvest was done and the livestock settled in for the winter, and leisure time was available, then the time seemed to drag and I thought July would never come.

THE DEFINING 1950s

Our wedding plans were seriously underway from the first day of 1950. Melville came to Maymont virtually every weekend, so she and Shirley could make arrangements for the "big day." I was busy caring for livestock and playing hockey, and then with seeding operations. The Imperial Bank was so pleased with Melville that they offered her the chance to resume her career in Maymont. She accepted, left the bank in Saskatoon in late June, and started with the bank in Maymont in mid-August. This was a big help to us because cash was very short on the farms at that time.

Finally, July 5 arrived and we had a wonderful wedding. The careful planning paid off with a flawless event. July 5 was a Wednesday, carefully chosen because of the "half-day holiday" to make it easier for local business people to attend. Our wedding time was 7 p.m. The first Wednesday in July was traditionally Sonningdale's Sports Day. I teased that I would be at the wedding if we didn't get to the finals. Mel countered by inviting the entire team, so there was no Maymont team at Sonningdale Sports Day in 1950! July 5 was a blistering hot day, but the Miller house, where the reception was held, was relatively cool.

What a wonderful start to our life together! We left the reception about 11 p.m. for our honeymoon, spending our wedding night in the Bessborough Hotel in Saskatoon, and then it was off to Waskesiu with an overnight stop in Prince Albert. After a 10-day stay it was back to Maymont to face the reality of farming and marriage. We hosted a wedding dance in early August in the Maymont Community Hall with a huge attendance.

Because of uncertain winter roads, Mel had to resign from the bank at the end of November. In the intervening period, she had obtained a driver's license and my parents had rented a house in Maymont and had moved. It wasn't long before our roads were impassable to wheel traffic, and we had to rely on horses for transportation to the town for curling, socializing, and supplies. Our brand new Chevy three-quarter-ton truck was idled until April of 1951.

Because of a snow storm on September 30, 1950, we were unable to complete the harvest until the following spring, but fortunately only 50 acres remained over-winter. I was ecstatically happy, but Mel, while happy, had never experienced the isolation of a snow-bound farm and found it, I believe, somewhat depressing. By contrast, I no longer was depressed by "farm isolation," because I had the woman I loved with me every day. I had discovered, in my teen years, that I was a "people person" and was happiest when busy with activities that involved others.

I returned to the farm from the University of Saskatchewan fully ingrained with the conviction that I had a duty of community service, a compulsion that would drive me for the rest of my life. So, Mel and I got involved in community matters.

Our marriage was a much greater challenge for Mel than for me. After all, I was returning to the house where I was born and it had always been my home, whereas she was not only moving into a "new" house, but also moving in with my parents. Such a situation has been the cause for much trauma in many marriages. However, Mother and Dad realized that things would be better if they moved to Maymont and they rented a house there. The next year, Dad bought the Orange Hall and converted it into an attractive and comfortable home, right on Main Street, which solved any isolation problems they had experienced.

The few months in which we lived together as two families were in the busiest time of the year, and it allowed them to see Melville in day-to-day situations. They were highly impressed with her work ethic, the respect she showed them, and what a good wife she was to their son. Their admiration for Mel increased every year as she demonstrated all the wonderful motherly traits as we raised our family. I know they concluded that I could not have done better in choosing a life partner.

We were, by far, the youngest couple in our neighbourhood and were looked upon for leadership, not only in farming, but also for things of community value. Along with Ron Nutbrown, who was single, we organized the electrification of our farms, which meant convincing our neighbours to sign up, and convincing SaskPower that our group should obtain some level of priority.

The rule for consideration was an average of 0.7 miles per outlet. Several trips to Saskatoon, meeting with Dan Dojack of SaskPower,

who proved to be a most wonderfully helpful individual, and by recon-
figuring the layout, allowed us to reach the goal, and our area got power
on July 31, 1952, an event that allowed for major lifestyle changes.

At about the same time as the power-line activity, I undertook to
organize the neighbourhood into a winter-roads project. The outcome was removal of brush and trees from roadsides, encouraging the
municipality to reshape the roads so they would not collect snow, and
the purchase of a snow plough. There were 12 miles of road and 13 farms
involved. We purchased a blower, and it was mounted on my tractor.
I was the main operator for six years. One winter I logged 100 hours.
It was also my responsibility to keep a log of snow removal hours and
to collect the money. For most of this time, all were happy with the
results. The roads got progressively better, and the need for the plough
decreased. No one wanted to go back to horse transportation in the
winter.

MARRIED LIFE

Mel and I developed a fine circle of friends in our local community and
in the Maymont District. We each played ball and curled; I also played
hockey. We each participated in United Church activities. I was active
in the Board of Trade, eventually as secretary, which meant I had the
responsibility to organize the annual Sports Day on Victoria Day. One
year, we had 91 ball teams entered in 8 or 10 events. Darkness overtook
us before completion.

On the farming front, we were increasing our beef cattle herd and I
was milking six cows, shipping cream, and raising two litters of pigs as
well as chickens, turkeys and, one year only, ducks—a real "mixed farm."
We slaughtered pigs, beef, and chickens for our own use, and sold eggs.
We had a large garden and generally provided for ourselves. We had
very little time during the day for frivolities; Mel had church meetings
and the odd social gathering of ladies. We upgraded our machinery and
no longer required hired help, except for the harvest when Mel hauled
grain as well as tended the children and did the household chores.

My experience at university inspired me to avail myself of the latest
farm management and production husbandry practices. In short, I was
aware that there were better farming techniques available, and I knew
how to access these improved methods. The University of Saskatchewan was a willing source of leading edge technology. So too was the

Saskatchewan Department of Agriculture, which had a major extension program, "The Agriculture Representative Service," best identified to farmers as "Ag Reps." This service covered the province, and there was an Ag Rep in relatively close proximity to every farm. The University Extension Division and the Ag Reps held numerous field-days and meetings to dispense relevant agricultural information and always responded to requests by telephone. This was a great service to those of us who wanted to be the best farmers we could be.

So, too, did the Canadian Broadcasting Corporation (CBC) convey helpful information, by way of "The Farm Broadcast" that ran from 1939 until 1965. This weekday program aired at 12:15 p.m. until 12:45 p.m., providing important information for active farmers. I hated to miss an edition. The program was produced in Toronto, hosted by Peter Whittal and Bob Knowles; it was then aired regionally from Ottawa, Winnipeg, Regina, and Edmonton, each region having a marvellous host: such as Dave Innes and Bob Knowles in Saskatchewan, George Atkins in Ottawa, Lionel Moore in Winnipeg, and Al Richardson in Edmonton. I arranged my noon hours so I could hear the broadcast.

A typical daily broadcast reported grain and livestock prices, and featured interviews with experts about the production of grain and livestock, and also pest control. These specialists shared timely tips that if heeded would enhance farm production. As well there was a 10-minute drama, "The Jacksons and their Neighbours," that related the story of two adjoining farms. There was the Davis family, who were on the cutting edge of farm operations, and "Dollar Dick" Jackson, who struggled to keep his farm viable. Jackson's nickname came from the fact that he never missed an auction sale and never paid more than $1.00 for any item. He also had a controlling housekeeper, Mrs. Sommerville, who was constantly urging him to do the right thing.

The Davis son, Bill, was married to Colleen Jackson, and that introduced an interesting dynamic to the relationship between these farm families. The whole scenario was totally believable, as every farm community could identify with the Davis family and "Dollar Dick." This marvellous capturing of rural life was authored by Mary Rogers Pattison, of Saskatoon, and was produced as a live broadcast in Winnipeg for the prairie region. It always contained subtle snippets of useful farm and household information. The Farm Broadcast was introduced and signed off each day by a theme song, "In an English Country Garden."

These three outside sources of farming information were invaluable to me as our farm grew and developed. Today, farmers still have access to outside help but in different formats. The University of Saskatchewan no longer has the Extension Division, but rather each College is responsible for dispensing information to the general public.

The University of Saskatchewan created the Centre for Continuing and Distance Education that arranges information seminars and teaches classes via satellite that beams the information right onto the computers of those who subscribe. Our daughter, Jill, is Program Manager of Agriculture and Horticulture Programs in this Centre and as such conducts or arranges many of its activities.

Suppliers of farm production products have agrologists on staff who provide technical information to the users who patronize them. CBC continues a daily one-hour program that covers a wide range of news items, as well as market reports and topical farm information; so, too, does CTV. The internet allows farmers to seek answers to their questions by way of 'googling'. By seeking and utilizing the latest in information technology the farm community continues to be well served.

We had a very good farming year in 1952, and our first child, Janice, was born on October 27. I recall vividly bringing this precious child home from the hospital in Saskatoon. I remember that evening, sitting on the chesterfield, holding her, and the tears pouring down my cheeks—what a wonderful emotional experience! I felt the same on

At home on the farm in 1965. Left to right: Janice, Mel, Jill (front and centre), Ted, and Joy.

April 2, 1954, when Joy was born, and again on December 23, 1957, when Jill arrived. These grandchildren drew my parents even closer to us, and they were willing helpers whenever needed. I cannot imagine anyone being happier than I, with this wonderful family, with numerous good friends, with a great community, and, even though I didn't recognize it at the time, the opportunity for leadership.

I really loved farming, especially working with cattle—even milking cows. I was not particularly fond of tractor work, although the urgency and perfection required in seeding and harvest created an excitement that was addictive, and demanded my presence on the tractor or combine.

Often, if the farm work fell behind, for whatever reason, I would hire someone to run the tractor to summer-fallow, and I would go fencing. This seemed to bewilder people, but it was simply because I enjoyed physical activity more than the monotonous task of driving the tractor. And the fence work had to be done correctly. I would often take one of the girls with me, and if they got tired, they simply slept. I really relished the company of the girls when I was doing a "safe" job around the farm.

Even before farming on my own, or in partnership with my father and brother, I was indoctrinated into the world of the Saskatchewan Wheat Pool. The cooperative philosophy fit easily into my own beliefs and personality. Working together was as natural as breathing and made for compatible relationships with my neighbours.

Ted checking on the grain crop at the farm in 1965.

One year, as I struggled to get the hay crop in, I purchased a nine-foot tractor-mounted mower. While that made quick work of the hay field, there was still the task of gathering the hay and storing it in a handy location. I hired the McIvors to bale the crop. They soon had it all done and I was impressed by the quality of the hay, and the ability to store all that I required for the winter in the shed attached to the barn.

The next year, Charlie Nutbrown, who lived half a mile away, and I purchased a baler together. It was adequate to handle the haying needs of each of our farms. In addition, I did a lot of custom baling and made good money, paying Charlie two cents per bale for his ownership share out of the 10 cents I collected.

CHAPTER 4
Early Involvement in the Saskatchewan Wheat Pool

EARLY INVOLVEMENT

As I mentioned earlier, my father was a lifelong supporter of the Saskatchewan Wheat Pool, and I followed in his footsteps, attending local meetings of the "Pool," and following its activities with great interest. When I was growing up, we always had a Wheat Pool calendar hanging in our kitchen. Part of the calendar was a map of Saskatchewan which showed every hamlet, village, town, and city in the province. It also showed the 16 geographic districts, and subdistricts within each district which provided farmer control of the Pool. Even as a child I thought that, one day, I wanted to be a part of that structure, and at more than just the committee level at Maymont. The obvious and only place for such involvement was as a delegate. Delegates were elected each year for a one-year term (this was later changed to every second year for two years).

In 1950, shortly after our marriage, Mel and I had gone to Saskatoon for the day and missed the local annual meeting of the Saskatchewan Wheat Pool. In my absence, I was elected to the Wheat Pool Committee. Thereafter, I attended the meetings faithfully, and became increasingly intrigued with the organization, lobbying Wheat Pool officials for a greater supply of rail cars for Maymont and working closely with the local agent to increase our market share. Colles Brehon, who farmed near the town, was our delegate, and besides asking questions about his reports, I explored the process, especially about the delegates' annual meeting, which always received a lot of radio coverage. I was unfailingly impressed by the speaking ability of the delegates. As well, Andrew Melrose, whose farm abutted ours, had been a delegate for many years, and because of my great esteem for him, I had further respect for the role of the delegate.

DELEGATE—AMID FAMILY LIFE

At this time, I enjoyed an excellent home environment. Jan and Joy were cute, alert, and good-natured children. Their mother always had them attractively dressed and as clean as could reasonably be expected. Mel quite often sewed clothes for them and on occasion would make look-alike mother/daughter dresses that they were all proud of. In 1957, Mel was pregnant again, and on December 23 we welcomed our third daughter, Jill, into our family.

Mel and I worked very hard with six quarters of farm land and some 40 head of cattle, and milked five or six cows. But we were young, and long hours meant nothing to us. We both were involved with the United Church, Mel with the Christian Girls in Training (CGIT) and the Women's' Group; and I with the Board of Stewards.

The 1957 harvest had gone very well, and mid-October found us with gorgeous fall weather and most of the field work completed. It was on a warm, sunny afternoon, when we were baling the last load of straw, that Don Sinclair, a Saskatchewan Wheat Pool fieldman, drove onto the field. Dad was there, as was our neighbour, Charlie Nutbrown. Don's request was straightforward. He had been asked by the Maymont committee to persuade me to run for the position of delegate for sub-district 1 of District 16 of the Saskatchewan Wheat Pool. There was to be a meeting that night, where the nomination papers would be signed. I was hesitant, unsure of my ability, and concerned that the time commitment might interfere with baseball and curling.

However, Dad and Charlie each encouraged me to run. I needed Dad's permission because we were in partnership. Meanwhile, Charlie said, "the work will get done," so that night I went to the meeting, signed the nomination papers, and four weeks later I was elected by acclamation. My life, and that of my family, would be dramatically changed by this event.

The Maymont community was genuinely supportive and I was proud to carry on my father's legacy of his tireless efforts in 1923–24 to start the Wheat Pool. While excited by the new undertaking, internally, I was frightened about what to expect and had doubts about my ability to do the job.

After my election, my first Wheat Pool meeting was with the other delegates of District 16, the main purpose of which was to elect a Director. This was a non-issue, since Jack Wesson was not only our Director, but also President of the Pool. However, the meeting allowed me to meet the other ten District 16 delegates, and it calmed some of my fears as I realized many of them were just like me, with a degree of uncertainty about our role. But all ten were friendly and dedicated to the Wheat Pool philosophy. During the next few months, I met with the seven Wheat Pool committees in my sub-district and found them very friendly and receptive. I also realized early on that many of the committee men knew more about the Wheat Pool than I did.

In March 1958 I attended a three-day seminar for new delegates in Regina. This was a real eye-opener and provided the base for the rest of my years in the Pool. It verifies a belief I hold, that proper education, early in the process, is invaluable, no matter what the undertaking. The course provided keen insight into the Saskatchewan Wheat Pool and created an understanding of the role of the delegate. In addition, I formed friendships that would last for decades—with Abel Toupin, Howard Tyler, Bill Hlushko, and Phil Rothery, among others. It was a pivotal experience.

Meanwhile, back on the farm, we were as busy as ever, trying to develop an operation to support our family and to ensure that when the time came, our three girls could each receive a post-secondary education. Every spare dollar was put into the farm, often meaning a sacrifice of other desires such as household improvements or personal items, except for one area. In 1955, I dug a trench by hand some 60 feet from the house to the well, eight to nine feet deep, and thus had water under pressure in the house, and in the late summer of 1958 I sold ten yearling heifers for a very good price. We used the revenue to install a flush toilet, water softener, water heater, and a sewage system. This complete water and sewage system was a great Christmas gift for all of us that year!

LEARNING THE ROPES

During my early days as a delegate I faced a fairly steep learning curve, but in large part this was met by going to local Wheat Pool committee meetings. My subdistrict ran from the Borden Bridge to Denholm, so there were six elevator points with seven committees, having what was called an inland committee at "Great Deer," some 20 or so miles northeast of Borden. My apprehension about meeting with the various committees was soon resolved.

Each committee had members of 20 or more years and they knew more about the Pool than I did, but they accepted me and were as helpful as could be expected. I met with each committee at least four times each year. It was a wonderful educational experience. Each community was different, and I learned to appreciate that uniqueness and soon learned to realize how such diversity broadened my understanding.

The delegates of District 16 would meet in North Battleford three or four times each year, for two days on each occasion. Jack Wesson was the director in attendance and conveyed all the latest happenings within the Pool. I didn't always know what he was talking about, but decided that patience was the best approach, and the real value of the meetings was to get to know my fellow delegates.

I was the youngest of our group at 30 years, three others were 40 or younger, one about 45, two were in their 40s, and the rest were 65 and older. The four eldest smoked cigars and, being a non-smoker, after a couple hours, I could hardly function. Our usual meeting place was in a regular room in the Auditorium Hotel, with the head of the table and the smokers at the window-end. At my second meeting, we younger ones got there first and put the head of the table near the door. Hence, we were able to open the window and we had first call on fresh air, which helped keep the smoke at the other end of the room.

As my friendship grew with the other delegates, so too did my comfort level with the job. In November 1958, I attended my first annual meeting in Regina. This was the first test of harmonizing the pressures of the farm with the demands of the delegate role. I was able to hire Pat Nutbrown and things went okay, but I found it very hard to be away from Mel and our three girls for a 10-day period.

Following the general annual meeting, there was the challenge of preparing a report for the seven local annual meetings. Don Sinclair, our marvellous field man, helped me with the first few. I was still terrified of public speaking, but necessity left me no choice and I eventually overcame my fear. Little did I realize that in 1985 I would be given a plaque by the Prairie Region of Toastmasters as Communicator of the Year.

As a delegate, I was involved 20 to 30 days a year, mostly at district meetings and the annual meeting, so it wasn't a drain on the farm, and I was able to hire help when I needed it, especially Wayne, Mel's brother, who worked for us on the off-days from Teachers' College.

The first day of the annual meeting was both eagerly anticipated and feared. I was frightened that I would be inadequate and everyone would know, but I also wanted the experience—and

John Henry "Jack" Wesson (1887–1965), served as the Wheat Pool president from 1937 until 1960. Courtesy of the Saskatchewan Archives Board, R-A15276(2).

what a magnificent experience it was! So much about the Saskatchewan Wheat Pool that I didn't know, and so many people in high-ranking positions, from government, industry, and universities. The acceptance and fellowship of the other delegates was wonderful; the discipline of the meeting was superb. I finally relaxed after I made my first trip to the microphone and realized that I hadn't bombed. People were always eager to support a first-time delegate and encouragement was offered freely.

The following two years saw me gain confidence as a delegate, as my understanding of the role improved and my knowledge of the Pool broadened. I developed a positive relationship with the seven Pool committees in my subdistrict. In October 1960, as I was closing in on three years as a delegate, our director, Jack Wesson, announced that he was resigning as president, director, and delegate of the Saskatchewan Wheat Pool. This caused quite a stir within District 16 because we would have the responsibility in December of selecting a new director for our district.

The chairman of District 16 delegates had been around for many years, as had two of the other delegates, and they were eager for the role. One other delegate of 10 years' service was also keen on the job. When we assembled in early December, the chairman asked who was prepared to seek the role of director, and went person-to-person around the table.

I declared that I would let my name stand. In the vote for director, I received seven of the eleven votes. Nobody else received more than one vote. I was really surprised, as were the others who contested the election. I left North Battleford that day, highly appreciative of the chance to serve, but frightened by the challenges of the new position.

I attended the December board of directors' meeting of the Saskatchewan Wheat Pool and participated in electing Charles Gibbings as president (he had been second vice-president), re-electing Tom Bobier as first vice-president, and Ted Boden as second vice-president. Also, I got to know the other directors a bit better, whom I had only seen during the delegates' annual meetings. However, a special meeting of the delegates of District 16 was called. When I attended, I discovered the meeting was to protest my election as director. The justification for overturning the election was that the district fieldman, Don Sinclair, had interfered in the election and had influenced several of the delegates to vote for me. Each delegate, in turn, denied that such was the case. Some expressed deep resentment at the thought that they could have been so influenced. Sinclair was vindicated, and my election upheld.

It was a rather rocky start to my tenure as a director of the Saskatchewan Wheat Pool, where four of the eleven delegates opposed the election of a delegate with only three years' experience—after all, they had anywhere from 10 to 30 years in the role. After a settling down period of a couple of years, I received strong support from the delegates and was never opposed for re-election as a director at any time during my 26-year tenure. The delegates appreciated the support I gave them in their own sub-districts and came to trust my judgement on most matters. I made it clear to them where I stood on all issues, especially on the controversial ones. I spent probably 15 to 20 days per year within District 16 in other than my own sub-district. This, while helping the other delegates, gave me an accurate feeling of the attitudes of members and allowed for the building of positive relationships with committee members and staff within the district.

The role of director did change our family life and the operation of our farm quite dramatically. In addition to the two-week annual meeting each year in Regina, I had to dedicate the third week of each month to a board meeting, also in Regina. This meant a lot of nights away from home and family. It also meant that Mel was called upon to do extra family duties and also farm-managing tasks. Somehow, with the aid of hired help—Wayne and our neighbour Pat—the farm work got done.

Also, following the September meeting, the board would go to Thunder Bay and Winnipeg to inspect facilities, interface with employees, and present service recognition awards. This helped board members to better understand the Saskatchewan Wheat Pool and provided a unique bonding opportunity among board members. In my early years as a director, these trips were travelled by railroad, as was our wintertime attendance at the annual meeting of the Canadian Federation of Agriculture. Fortunately, I was able to include Mel in most of these ventures. One year, Jill, at the age of five years, accompanied us on a rail trip to Ottawa—she was the "hit" of the group!

We had established a routine that took into account my frequent absences and in January we would move to my mother's home in Maymont, when she would go to live with my sisters in Rosetown. This simplified school and event attendance for Mel and the girls. Still, there were pressures, and I am not sure how long we could have continued this way of living. In 1964, I slept away from home on over 200 nights.

DIRECTOR—THE FIRST SIX YEARS

The first six years I spent as a director was a huge personal development period for me, with new lessons learned every day. First, I had to establish a positive relationship with the delegates in District 16, especially so with those who opposed my election, and I had to prove to the others that their support of me was well-placed. This I was able to do on each count, as I described earlier. But of greater challenge was the new culture of "board dynamics." The Pool board was large—comprised of 16 competent and dedicated men.

There was a strict discipline at board meetings. The meetings were long and wide-ranging. I was relatively inexperienced, having only been a delegate for three years, and I was thrown in with others who had a range of serving from seven to 30 years. To compound the situation, I was 33 years old; the next closest was Jack Stilborn at 38 years, then Ted Boden and Charlie Gibbings at 43 years. There were a couple of directors in their early 50s, and then the rest ranged from 60 to 75 years.

I was accepted enthusiastically by all of them, and they were patient in dealing with my lack of knowledge. Some told me years later, after I had been president for some time, that they sensed a latent leadership quality in me and were grateful to have a "young person" on the board. A board of directors takes on a definite personality, just as does an individual. In fact, a board is ineffective until it develops a personality that is a product of the individual contributions of its members.

There were numerous opportunities to bond with other directors— a five-day board meeting each month, with evenings spent together in hotel rooms, the trips to various meetings on the train, and meetings of the standing committees. Jack Stilborn and I, perhaps because of our proximity of age, became very close, and I had a positive relationship with all of the directors. Many of them visited us on the farm at Maymont. Mel was a huge asset in this process as a hostess, and also with her presence at board gatherings and when she joined us on some of the trips.

When I first became a delegate, none of the wives of delegates in District 16 even came to the annual meeting social functions. The second such event, I encouraged Mel to come, and with my parents looking after our three daughters, she got to enjoy a new experience. Before long, several District 16 delegates' wives were attending regularly. So, too, at the director level, Mel attended functions whenever it was possible and before long, other directors' wives were attending. Her pleasant personality was a real friendship-maker and set a trend that would support me the rest of my life in numerous undertakings.

The first annual meeting Mel attended was in 1959. The main social event was a "tea" on the first Friday of the meeting. This was a semi-formal event where the wife of each of the president and vice-presidents would welcome guests at the door. After arriving in Regina, I met her for lunch and she indicated a reluctance to go to the function; she felt out of place and thus uncomfortable. I persuaded her to give it a try, even though there was no one else from our district to accompany her. She went, and a lady named Ethel Toupin, whose husband Abel once served District 16 as a field man, seemed to sense her anxiety and within minutes, put Mel at ease and introduced her around. Mel now often laughs about how scared she was, little realizing that some years later, as the president's wife, she would be the main hostess. Remembering her own first experience, she endeavoured to put all first-time attendees at ease.

My first day at a board meeting as a director was Charlie Gibbings' first as president. Charlie had a remarkable grasp of the Saskatchewan Wheat Pool and a great concept of the potential for this farmer-owned co-operative. He was surrounded by completely dedicated directors, but most saw their jobs more as caretakers than as developers. I do not wish to disparage these men because their attitudes were related to age, and

Charles William Gibbings (1916-2009), served as the Wheat Pool president from 1960 until 1969. Courtesy of the Saskatchewan Archives Board, R-A15239.

they had all made remarkable contributions to the Saskatchewan Pool. However, the Saskatchewan Wheat Pool was marking time. We were truly a "management board" and as such, it was difficult to move the organization forward as a unit, when valuable time had to be used to harmonize between the various divisions.

Charlie recognized that changes had to be made, and while being less precise in what was needed, I sensed that there were many unrealized opportunities. Charlie also knew that you cannot turn an ocean-liner onto another course quickly, or indeed, you cannot turn it at all if it is standing still. He knew that a gradual change was probably best, and all that could be achieved.

Very quietly, and without me realizing it, Charlie cultivated me as a colleague and as an individual. Whenever Mel and I were in Regina over the weekend, we were invited to his house for Sunday dinner. He introduced me to golf on a solitary weekend—my first golf experience. At the board level, he quietly directed responsibilities (opportunities) my way, often combining me with "progressive" counselling directors.

49

For instance, for some reason the manager of the flour mill and the manager of the vegetable oil crushing plant did not speak to each other. Charlie created an industrial division committee and teamed me with Louis Boulieu, vice-president and William McKenzie Ross, another director. What a learning opportunity as we restored harmony in the two operations that begged to be under one manager. We never got the individuals to speak together, but a few years later, the two functions combined as one.

MAJOR DECISIONS

There were several decisions we made in my early years as a director that were major in nature and in implication. The first major decision, after much agonizing by the board and a management study, was to purchase a main frame computer for head office. Consultants were utilized and the ability of the staff was assessed. Everything seemed positive and the cost-benefit analysis also seemed favourable. On the recommendation of management, the board approved the purchase. However, there was a massive negative reaction from the staff and from the Grain Services Unions since part of the justification for the purchase was a saving in labour costs.

Finally, Charlie Gibbings as president sent a letter to each head office employee saying that no employee would lose his or her job because of the computer. This seemed to calm the outcry and so the preparation for the installation and staff training began. The outcome, of course, was that far from reducing jobs, many positions were added, and our ability to do more things was greatly enhanced. It was a sound decision and very basic in preparing for the future growth of the Saskatchewan Wheat Pool. The board's decision was upheld and the union was happy as its numbers increased.

The second major decision took place in the early 1960s when we were faced with the need to offer greater services to our members. We were doing a good job in grain and livestock marketing but did not offer products to support the production of those commodities. Charlie proposed that we sell farm supplies. Several of the other directors and I supported the suggestion as a way to round out our service to producers. However, several other directors felt that we would be invading the territory of Federated Co-op Limited by going into this new area.

The same sort of debate took place at the annual meeting. However, the board made the decision to proceed with selling farm supplies. The Country Elevator Division would handle the products, mostly weed spray and fertilizer. This caused protests from some elevator agents who saw the venture as an added burden to their jobs. Other agents saw it as a way of getting more farmers through the door. There were violent protests from local co-ops and also from Federated Co-op. In the rural areas, our very best customers were also strong co-op supporters; many were on the board of the local co-op and some were on the board of Federated Co-op itself.

I felt that by providing more outlets, the whole co-operative supply system would benefit. In many areas, there was not a co-op presence; for instance, while still on the farm at Maymont, several of us had worked with our elevator agent to sell a carload of fertilizer that we originated through the North Battleford Co-op. This worked well. Farmers got their product from Maymont and the co-op got extra volume without any effort. However, the next spring leading up to seeding, the co-op did not even contact us. We had spent considerable time the year before and were not prepared to do so again. The result was that a local dealer brought in fertilizer and sold it easily. I hauled mine in from North Battleford at extra cost, and the non-caring co-op lost some business.

This debate over our entrance into farm supplies went on for years. When challenged about it at country meetings, I simply replied, "Buy where you wish, as long as it is co-op product." The Saskatchewan Wheat Pool presence in the retail field enhanced the operations of Western Co-operative Fertilizers Ltd. and also of Interprovincial Co-operators Chemical Company.

At the outset, we did not have retailing expertise within the Saskatchewan Wheat Pool. Hank Brown, a very personable employee of the Country Elevator Division, assumed the role as head of our farm supply business. It was a slow start, but the momentum picked up and it soon became apparent that the business was outgrowing the Elevator Division. The board made the decision to establish the "Farm Service Division." We enticed Roy McKenzie, director of the Plant Industry Branch, Department of Agriculture, Government of Saskatchewan, to switch careers and to join the Saskatchewan Wheat Pool as manager of our new division.

It proved to be a marriage of mutual benefit. Roy worked closely with Hank Brown and recruited the brightest agriculture grads from the University of Saskatchewan. Roy felt it was necessary to "service what you sell." To do so required technical knowledge, and over the next 20 years, the Farm Service Division developed a marvellous system of centres serving every area of Saskatchewan. The centres sold directly, but for the most part, supported the sales made by elevator agents. From its start in the early 1960s, the Division moved from several hundred thousand dollars in sales to $200 million in the 1980s.

Roy had been brought into the Saskatchewan Wheat Pool in 1964 to advise on farm supply sales. In 1965, he was made manager of the Division; and in that year, he reported sales of $5 million. In his retirement report in 1982, he advised the annual meeting that sales that year had reached $155 million. Some members felt that because the Saskatchewan Wheat Pool was in the business, products would be at a lower price than at any other outlet. However, the Division quite correctly operated on a margin that would allow full service of sales and also expansion of facilities.

On the recommendation of Roy, and on follow-up investigation by Ira Mumford, general manager of the Saskatchewan Wheat Pool, and with subsequent approval by the board of directors, the Saskatchewan Wheat Pool purchased three quarter sections of land near Watrous. We then undertook field trials to test different grain varieties and the effectiveness chemicals had on common weed species. We invited outside companies in to test their products on a confidential basis. We worked closely with the University of Saskatchewan, so we complemented each other. Grain breeding programs were undertaken and this attracted plant breeders, and at the same time, helped to develop our staff. It was a highly successful operation at a minimal cost. The farm developed a new wheat variety, "McKenzie," which was the first privately developed wheat variety in Canada. It also signified to our members that, "we are proactive in assessing your needs and in moving to meet your needs." The new industry contacts were also invaluable. It complemented our commercial operations and also our public policy thrust.

In his last report to delegates, Roy noted that he had appeared before them at annual meetings 18 times. He said that, in total, he had been asked 540 questions, listened to as many comments or suggestions, and heard debate on more than 600 resolutions concerning farm supplies.

Thus, a controversial decision in the 1960s had turned into a shining example of success because skilful management responded to a need.

The third major decision was to build a new grain terminal in Vancouver. The Saskatchewan Wheat Pool was leasing a rather small, inefficient terminal that did not lend itself to upgrading. For years, the largest portion of Saskatchewan-grown grains left for export through Thunder Bay. The Saskatchewan Wheat Pool had the most storage space in that port and had seven terminals in total, most of which were operationally excellent.

Thunder Bay had several severe drawbacks, however. It relied on ships to move the grain to destinations that were either another country or grain transfer plants on the lower St. Lawrence. There were size limitations on the ships because they had to fit into the lock system leading to the St. Lawrence River. As well, a year's business had to be crowded into eight months because cold weather prevented operation on the Great Lakes from about December 20 to April 10 of each year. This created inefficient operations of the shipping elevators and caused overseas customer problems in organizing their receipt of grain shipments. With so many elements at play, there were many possible means of disruption.

In the 1960s, there was a discernible shift in markets from Europe to Asia. As it recovered from the Second World War, Europe was becoming self-reliant on grain needs, while countries like Japan and China were seeking larger quantities of Canadian grains. Therefore, we had to adjust our Thunder Bay focus to the Pacific. Clearly, as the largest originator of grain, the Saskatchewan Wheat Pool needed the facilities to service this expanding opportunity. Investigation began on the most appropriate site for a new, large terminal on Pacific tide water.

The main point of interest was Boundary Bay, where there was deep water available to accommodate ocean vessels and with lots of room for railway trackage to accommodate rail movement from the prairies. It seemed too good to be true, and it was. Engineers could not find bed rock upon which to rest the terminal. There appeared to be about 300 feet of silt, so we abandoned that potential site and reluctantly moved onto the north shore of Burrard Inlet. The site was quite satisfactory but congested for both vessels and rail cars. It was a difficult decision for the Saskatchewan Wheat Pool Board because a satisfactory return on investment was by no means assured. However, the need was evident, and so the board made the decision to build a new state-of-the-art grain terminal. The board worries were for naught. Markets were large and an efficient plant provided a positive return.

This is a prime example of how carefully the board of directors and management handled the financial resources that really belonged to the Saskatchewan Wheat Pool members. This conservative philosophy may have cost expansion opportunities, but it also kept the Saskatchewan Wheat Pool in a viable financial position at all times.

THE HOG MARKETING BOARD

In 1964, the National Farmers' Union asked the Saskatchewan Wheat Pool to appoint someone to the provisional board of a Hog Marketing Board. I am sure that Charlie Gibbings encouraged my appointment. Our board supported the concept but there were many long-standing irritations with the National Farmers' Union that caused our longer-serving board members to resist the association. I carried no such baggage and so most directors were pleased to have someone in the role as long as it wasn't them—a sort of election by default.

This turned out to be another growth experience. I attended dozens of meetings around the province promoting the potential board, and at every one of the meetings, encountered strong vocalized resistance. However, I became even more convinced that such a board would be a real asset to hog producers. These meetings were my first experience of personal attacks, including the most difficult one to defend: "You are not even a producer in your own right." This taught me the valuable lesson that personal attacks, while not pleasant, are often the result of frustration felt by those who had lost a logical argument, and felt that discrediting the carrier was their last chance to make their point.

During the spring of 1965, I attended a national conference on hog marketing in Montreal. This was the first time I met Glen Flaten, a Regina hog producer and a strong opponent of marketing boards. Glen had been attacked by the National Farmers' Union because he had a marketing arrangement with a packing plant. We spent time discussing the situation, and with presentations made at the meeting, Glen sensed the value of single-desk selling. As seat mates on the way home, he said, "Perhaps I am on the wrong side of this issue." Glen never came out in support of the board because he was still smarting from the earlier attacks by the National Farmers' Union. However, the Montreal meeting did convince him of the value of farmers taking charge of their own industry. He went on, some years later, to be a very effective president of the Canadian Federation of Agriculture and, even later, to be equally competent as president of the International Federation of Agricultural Producers and a strong proponent of marketing boards.

Interestingly, a few years later, a hog producer/trucker from Fielding apologized to me for all the things he had said about me. "I knew they weren't true, but I was afraid I would lose my trucking business, and it is my main source of income." A few years later, in 1972, a Hog Marketing Board was created and still operates today. Perhaps our early efforts were not in vain. I came out of the exercise not unscathed, but with a better understanding of human reactions and better prepared to withstand personal attacks.

One Saturday, I was returning to Maymont from having spent the week attending meetings promoting the Hog Marketing Board in southern Saskatchewan. I knew there was a farmers' bonspiel in Radisson, so I stopped at the curling rink for a cup of coffee and to check things out. I was immediately confronted by two committee members from Fielding, a village next to Maymont, and only six miles from our farm.

The committee was very volatile and they loved to argue among themselves, with the members lining up on different occasions in different camps. It was the norm that after a committee meeting was over, a discussion or "debate" would go on about some topic for another two hours. I often sat in on this ritual. I was accepted; if someone not accepted took sides then the whole group would turn on that person. They were nothing, if not blunt! I soon found from the two who confronted me that they were upset by me being on the Provisional Hog Marketing Board and the fact that I was actively promoting it. As it turned out, most of the committee members were in the curling rink, so we got a room and held a meeting. Their point was that they were opposed to the Marketing Board and because I was in favour, I should resign as their elected delegate.

My response was two-fold: (1) "I represent seven committees as delegate, and therefore, the other committees needed to have input into the demand on me to resign"; and (2) "You have elected me to represent you in the functioning of the Saskatchewan Wheat Pool and, therefore, you delegate me to use my best judgement in making decisions.

You know that I consult with you in advance when there is time to do so." I gave several examples of just that. I continued, "I will not resign because I have used my best judgement in this case. Your only recourse is to nominate someone to run against me in the delegate elections next November. That is your right and I will not take offense from it if you do so ... but I will try to beat the hell out of your candidate in the election."

This seemed to settle the issue; although they were not in favour of a Hog Marketing Board, they no longer felt that I betrayed them. Through all my next 19 or so years, the Fielding group were my strongest supporters. They confronted one openly and bluntly, but they also respected a blunt and honest reply.

Actually, I was rather surprised by the way I articulated the ground rules for representative democracy. They learned a lesson and I grew in stature and learned the value of being honest and direct. This experience also helped my confidence in handling unexpected situations. This, combined with the ability to handle country meetings with an increasing competence provided stability but not cockiness to my demeanour.

When I first became director, the task of attending the spring meetings in each sub-district was very scary indeed. I was almost terrified of each event; however, I soon learned that those attending were very supportive and only a few people chose to be destructive in their remarks. They sensed my lack of experience but seemed to be pleased to help this young director "understand better." I soon came to know many of the sub-district people by their first names. Knowing District 16 was also enhanced by assuming responsibility for reporting to the annual meetings of local points. The fieldman Mac Lambie and I would each take half of the local annual meetings. This allowed me to know agents and committee men, but I did not realize at the time that it impeded the development of the individual delegate.

BUSINESS IN THE SUB-CONTINENT

Another event in which I was privileged to participate occurred when, once again, the National Farmers' Union requested the Saskatchewan Wheat Pool to send two individuals on a National Farmers' Union-organized trip to India. The purpose of the trip was to evaluate agriculture and to examine Canadian programs in that country. The board selected me and the assistant corporate secretary, Jim Wright, to go on this February 1966 trip.

Jim and I had been classmates at the School of Agriculture, so we were already compatible. Jim and I forged an even stronger bond that lasts to this day. Jim, a few years older, had experience overseas in World War II. As a prelude to that trip, Mel and I went to New York three days prior to departure, and Jim joined us there a day or so later. We had a good time in New York and put Mel on a plane back to Toronto a few hours before we were to embark on an Air India flight to London, England. The flight was delayed and I had my first and last encounter with martinis. They were very smooth, but I paid the price on our flight and vowed to never drink one again. Promise kept.

Our three weeks in India were astounding. Never had I imagined such poverty. Never had I witnessed such disparity of wealth. Never had I heard such laughter on the streets. Never had I seen cattle running on city streets. Never had I seen such wonders as the Taj Mahal, the Red Fort, the garden at Hyderabad, or the tea plantations in Madras Province.

The people were friendly, the weather extremely hot, the food a challenge on occasion, and the dreaded Indian two-step (you don't dare venture further than two steps away from a toilet) visited everyone in our group. Paul Babey, the leader of our group, and at that time head of the Alberta wing of the National Farmers' Union, was excellent, and he alone escaped stomach upset. Perhaps his daily commitment to a good stiff drink (or two) of rye proved to be preventative medicine. Paul and I became good friends and remained so during his tenure as president of the National Farmers' Union, and later in his executive role with the Farm Credit Corporation. Jim and I returned to Canada via Japan and Hawaii.

In March 1966 I was faced, as every year, with a round of sub-district banquet conferences. In District 16 at that time, it meant 11 such meetings where I, as director, would report to the committee men in each sub-district. I had a good camera and at the conferences, I showed some 80 slides that were set up to tell the story of my trip and to expand the horizon for the conference participants. This was tremendously popular and I was soon called upon to show them in many places around the province. I must have, over the next few years, shown that set of slides over 100 times.

ELECTION BUSINESS AT HOME

Seeding time and all the preparations were upon us, as were the April, May, and June board meetings. I had been a director for over six years and had grown dramatically in the role. I was well-established in relationships with other directors and with head office staff, and I had a good rapport throughout District 16.

In 1966, First Vice-President Louis Boilleau died, and to my surprise, I was elected to replace him. Other directors had persuaded me to run in the ensuing election. However, we had expected Second Vice-President Ted Boden to advance, and I would then contest his vacated position. Lo and behold, after the first ballot, there was a tie for the position of first vice-president and the chairman ordered a re-vote. On the first ballot, I had voted for Ted Boden, the second vice-president, so I then voted for myself and was so elected. This meant that I must move to Regina immediately.

This change came at a convenient time; that is, prior to the start of a new school year, so we were able to settle the girls into their respective schools. To leave Maymont was a traumatic experience, as it had essentially been our only home, and now we were moving to a city some distance away, leaving behind lifetime friends and familiar surroundings. For the girls, who each had many close friends, it was overwhelming, and only the prospect of a new adventure allowed them to carry on. The Maymont people were wonderfully supportive and somewhat proud that one of theirs was making a name for himself and the community. The community held a farewell evening for us. The community hall was jammed full and the program a delightful blend of humour, music, compliments, and best wishes for success. On Labour Day weekend 1966, we left Maymont and settled into our new house in Regina.

THE OFFICE YEARS

I was elected on July 16, 1966, to the position of first vice-president, and the following Monday I arrived in the office with nothing more than a pen. The first task was to understand office routine. I felt like a fish out of water, but Rose May, the secretary to the president and two vice-presidents, taught me how to use the dictaphone and general office procedures. There were some documents to be read and I seized every minute Charlie Gibbings and Ted Boden had to spare, in order to learn what was expected of me. I made arrangements for a hotel room, and since I was now on a salary, I could not claim hotel or meals as an expense. Interestingly, there was considerable discussion at the board meeting about what my salary should be; it was finally settled at $13,000 per year.

Suffice it to say, I learned the office procedures and needed to get into carrying my load as an executive member. The workload was split as follows: Ted Boden would remain heavily involved in the public policy for agriculture activities. I would assume all activities related to co-operatives and Charlie as president interacted with management of the various Saskatchewan Wheat Pool divisions, and was the chief spokesperson for the company. Charlie also interacted with governments and was the final decision-maker for the entire organization. The elected officials shared the attendance at country meetings associated with elevator openings, special member meetings, and on occasion, district meetings of delegates.

In the co-operative area, I was soon on the board of the Saskatchewan Co-operative Association and a couple of years later, president. I was on the board of the Co-operative College located in Sutherland, and also on the board of the Co-operative Union of Canada and on its executive. The Co-operative College was a facility that co-ops used to train their elected members and their staff. It also had a clientele from developing countries and made a huge impact on the economy of those countries by training people on how to establish and effectively use co-operatives for productive enterprises. Co-op College was revered in many countries and, by the same token, the Saskatchewan Wheat Pool was mentioned as a model of a successful co-operative.

The college did much to help cohesion between co-operatives because delegates, directors, and management would assemble there for a particular course, and as a consequence, would better understand each other. As our co-operatives became more successful, then the decisions were made to do in-house training; in retrospect, this was an unwise decision because of the lack of inter-co-operative exposure. I relished those days because of the nature of the work and the dedication of a competent staff.

The Saskatchewan Wheat Pool was comprised of seven separate companies, the role of overseeing this structure was a board responsibility. A board is ill equipped for such a role and there had been talk for several years of appointing a general manager to assume these duties. However, each time it was discussed at a board meeting, it was rejected because the board of directors felt they would lose control. Charlie Gibbings was convinced it was the way to go and I was ready to support him because I wasn't convinced that lack of control was an issue. We had, on several occasions, used the Management Consulting branch of Touche Ross for various purposes. We sought their advice and, not surprisingly, they recommended strongly that a general manager should be appointed. Touche Ross presented us with terms of reference for the position. The board had approved the study and because we were well-prepared with our presentation of the Touche Ross report, it was approved.

A search was conducted and after a few weeks, we were able to recommend to the board that Ira K. Mumford, presently the corporate secretary, be appointed as general manager. This was adopted by the board. One of the worries of board members had been that the new person would not understand the Saskatchewan Wheat Pool. The recommendation of Mumford alleviated that fear. Ira assumed the position and had a brilliant career as general manager. This exercise, once again, demonstrated the value of patience, proper preparation, and presentation.

During my second year as first vice-president, I was gaining confidence, and office procedures were no longer a challenge. We had a very harmonious relationship within the presidential group; there were new directors on the board with an attitude conducive to moving the company forward; things were good! One day, Charlie walked into my office and said, "I had committed to attend a meeting in Purdue two days from now, but it is essential that I be in Ottawa that day. Will you attend in my place? The community is upset about our support for ending the feed grain subsidy to Eastern Canada."

I said, "Yes, I can go, but I am sure it is you they want!" Ian Bickle, director of communications, agreed to go with me. I contacted the secretary of the Kinley Wheat Pool Committee, who was organizing the meeting, and told him the situation. He wasn't pleased but gave me a few details, time and exact location. The next day he phoned and invited Ian and me to come to their farm for supper and then to go with him to the meeting. We had a very nice meal and a pleasant chat, and about 7:30 p.m. he said that perhaps we should leave, as it was seven miles to the meeting. We arrived at the hall about 15 minutes later, to find every chair occupied, and people standing. People were not in a good mood, the meeting had been advertised as 7:30 p.m.; I had been told 8:00 p.m. Not only did they have to settle for a substitute, but the substitute was late!

Before I could catch my breath, I was invited on stage to a lonely chair, while the chairperson, who proved to be very capable, explained the ground rules for the evening. I was to speak for not more than 20 minutes; there would be questions and comments from the audience; then resolutions could be presented and debated. I spoke for about 10 minutes, outlining the reasons why the Canadian Federation of Agriculture had supported the removal of the Feed Grain Subsidy. The paper had quite correctly reported that Saskatchewan Wheat Pool delegates Joe Harrison and Bob Fulton had voted in favour; therefore, in the minds of the audience, the Saskatchewan Pool was responsible for lowering the price of feed grains—an accurate analysis but one which ignored the fact that the livestock industry in the prairies would likewise benefit over Eastern Canada.

A lively question-and-answer period followed, and I was soundly chastised. However, I stood my ground, kept my cool, and was honest. If I didn't know the answer, I said so and offered to relay the information sought by the questioner. A resolution was put to the meeting that condemned the Saskatchewan Wheat Pool and instructed that the Feed Grain Policy be reinstated. The chairman turned to me and said, "In this community, everyone is accountable, so I will call the vote. Everyone in favour remain seated, and/or slump down." I took two quick steps across the stage, sat on the chair, and said, "Look! Only one person standing—the Chairman!" The place broke into long laughter! The resolution was declared "passed" and we went back to the question-and-answer session. After one and a half hours, the meeting came to a close. I thanked everyone for attending, and by way of their participation, keeping the Saskatchewan Wheat Pool strong. As I left the stage, I received a standing ovation from all but a few. The chairman did not allow abusive remarks and I was accepted, even though my answers were not what they wanted to hear. The lesson, "Be honest, be forthright, do not react by being abusive, and show respect for those who are challenging, remember they have a right to do so, they own the co-operative" was well-learned.

BANFF SCHOOL OF ADVANCED MANAGEMENT

The Saskatchewan Wheat Pool, for several years, had been a strong supporter of the Banff School of Advanced Management and had sent one or two executives each year for the six-week course. In 1967, we enrolled Charlie Leask, manager, livestock division; Ian Bickle, manager, communications division; and me, as first vice-president. The course started in early February and ended mid-March with a five-day break at mid-point.

The curriculum was full. Classes started at 7:30 a.m. and went until noon, then a study period for two hours, a recreational break until 5:30 p.m., when dinner was available. Then structured study groups were in place from 7:00 to 9:30 p.m., followed by a social time for as long as you wished. Classes were held on Saturdays but adjourned at noon; we were required to be at study sessions on Sunday nights. This was a very intensive schedule, but the richness of the material, the competency of the instructors, combined with the practical knowledge of the student participants created a productive learning environment. I soaked up what was offered and it was useful to me for my entire career. I remained friends with the principal, Bob Willson for many years and used him in an advisory role on occasion for Saskatchewan Wheat Pool development functions.

COMMONWEALTH STUDY CONFERENCE

In 1962, Ira Mumford attended the Duke of Edinburgh's Commonwealth Study Conference, held in Canada. When the same event was scheduled for Australia in 1968, it was determined that I should apply to attend. The conference tried to keep participants within the 40-year-old age maximum. I was 40 at time of application and 41 when the event took place, from mid-May to mid-June. I was accepted and asked to be the chair/recorder for one of the themes of the conference. The event was comprised of 300 individuals from the Commonwealth countries. There were opening plenary sessions attended and inspired by the Duke himself. The 300 participants then split into about 30 groups and spread out across the country to talk to businesses, individuals, governments, and organizations about the effect of industrialization and re-industrialization upon individuals and communities.

My group went to a region in Victoria State, Australia, which encompassed South Melbourne and Geelong. We visited factories and homes, sporting events and local governments, charitable organizations, and met with individuals. Our group leader, O'Keefe, a mining executive from Broken Hill, Australia, was excellent. There were numerous receptions, all in the same format—a large room devoid of furniture. We were served a beer and picked food, displayed in great abundance, from a buffet table.

The interplay within the group was the real essence of the conference and most stimulating; our days were long and we were either sleeping (six hours), eating, drinking, interviewing, or debating. After two weeks of such activities, we assembled in Melbourne for the five-day plenary session wind-up.

The first two days of the final five were used to prepare reports; our study group elected our chair and two others to do so. Because I had agreed to develop one of the themes for the conference, I had to pull from all the study groups their feelings on the matter, condense it into a report and personally, at a plenary session, present it. This was challenging and used up all of the time prior to the wrap-up function. While others were out enjoying the best offerings of Melbourne, I was condensing feelings and framing a report in official-sounding language for presentation to the Duke.

There is no doubt that this conference, half a world away, expanded my horizons and at the same time, increased my awareness that, as a business, we must be conscious of the effect our decisions can have on others—especially on those who are powerless or not well-equipped to adjust or respond.

There was "a dream come true" effect for our family. Mel had, at the age of 12, started to correspond with an Australian girl one year younger than she, who lived in Ingham, North Queensland. They had faithfully kept in touch for almost 28 years. When I knew the dates of the conference, Mel and I set in motion plans that would allow Mel and the girls to fly to Australia some 10 days after the start of the conference. This happened as planned, and when I joined them in Ingham, it was like I was joining a long-established family. I met Vera and Sam Spina for the first time in person, but of course "knew" them from being privy to years of correspondence. Their children fit nicely with ours. John then Jan, Rosemary then Joy and Jill. Mel said that by the time they reached Ingham from the Townsville airport, the bonding process had started and would continue uninhibited for the rest of our stay and on into the future.

Vera visited us in Regina on five occasions, and Mel returned to Ingham four times (I was with her on three of these occasions). Sam, unfortunately, was accidentally electrocuted at his farm workshop a few years after our first visit, but his lessons of fishing and sugar cane farming live on in our memories. This venture was a nice decompression event following an intense conference and I learned even more about the Australian culture. It was wonderful to see Mel and Vera cement a friendship that had started years earlier. Life in the tropics is different and we all learned from it. Even though Vera, too, is now gone, we continue a friendship by e-mail with her family. Joy and her family visited Ingham as part of an Australian camping venture; it served to further confirm a positive relationship.

Then it was home to pick up the stack of duties deferred by six weeks in Banff and five weeks in Australia. This meant long but satisfying days. The delegates in District 16 were understanding and I did not want to disappoint them, so I worked hard to meet whatever commitments they wanted from me.

As 1969 dawned, I had been a delegate for over 11 years, and a director for over eight years. There had been significant changes in District 16. As I moved between the various sub-districts, the committee members were noticeably younger, probably in response to a younger director. This was also reflected among the delegates.

GROWING IN CONFIDENCE
Probably the most significant change in the period was in my own growth, competence, and confidence. I learned on the job, an apprenticeship-type learning from such things as the challenge to my election as a director, Saskatchewan Wheat Pool Board meetings, board committees—livestock and industrial, district delegate meetings, spring banquet conferences, meetings in other districts (these presented challenges in interpretation as compared to familiar District 16 meetings), the hog marketing campaign, vice-presidential duties, co-op association involvement, experience in debating issues around meetings of the Canadian Federation of Agriculture and the International Federation of Agriculture Producers (IFAP), and attendance at meetings such as that in Perdue, described earlier.

These varied experiences prepared me for a rather startling event that rocked the Saskatchewan Wheat Pool, and indeed, the whole grain

Saskatchewan Wheat Pool delegates from District 16, 1985. Back row, left to right: Robert Iverson, Rudy Jurke, John Simmonds, John Ehnes, and L. Larson. Front row, left to right: Ted Turner, Russ Arnold, Dennis Vanderhaegen, and Walter Campbell.

industry, when Charlie Gibbings resigned from the Saskatchewan Wheat Pool to become a commissioner of the Canadian Wheat Board.

Our board meeting was scheduled to be held in Saskatoon on Monday, June 16, 1969. Two nights previously, Charlie phoned to say he would not attend the meeting and gave the reason why. It was with mixed emotions that our board, especially I, received the news—on the one hand, not knowing how we could fill the vacancy he would leave, and on the other hand, pleased and confident of the positive contribution he would make to the Canadian Wheat Board.

As first vice-president, I was in the chair as the board meeting opened. After the usual announcements and minutes, the first order of business was to elect a president. I stood for election and was acclaimed on the first ballot. My vacated position was filled by Ted Boden and his previous position of second vice-president was filled by Don

Ted Turner, upon becoming president of the Saskatchewan Wheat Pool in 1969.

Lockwood from Davidson, the director of District 10. Don and I had been friends for many years. I would often pick him up on the way to board meetings and we roomed together most of the time. So, our executive group would remain harmonious and we found a way to pick up the slack from Charlie's departure. Charlie was so well-respected within the Saskatchewan Wheat Pool and in all our associated companies that it was a daunting task to fill the void of his departure; the brunt of it fell to me because I was now the chief spokesman and decision-maker. I was completely supported by Ted and Don and to the highest level by the delegates in District 16. Key management people like Jim Wright and Ira Mumford were totally supportive.

CHAPTER 5
President of the Saskatchewan Wheat Pool

PRESIDENTIAL DUTIES

In essence, the president of the Saskatchewan Wheat Pool was the chief governance officer, aided by the corporate secretary and the country organization department. The very strength of the Pool was in its democratic structure. Because the ownership base was so broad—70,000 shareholder members—it was not feasible to have decisions vetted by members. Therefore, an effective administrative structure was put in place. The province was divided into 16 geographical areas, numbered 1 to 16. Each district in turn was portioned into 9 to 11 sub-districts. Each sub-district elected a representative—a delegate—in the beginning to serve for one year, and later for two years. The delegates within the district elected a director on the same time basis. The 16 directors so elected comprised the board of directors of the Saskatchewan Wheat Pool and were the ultimate decision-making body. The board, in turn, appointed a general manager/chief executive officer (CEO), who was responsible for the commercial operations. They also appointed a corporate secretary who was responsible for corporate records, legal counsel, and for member services. This was "representative democracy" in its truest form.

The president was ultimately responsible for safeguarding the integrity of the system and for acting as the board's representative with the

Saskatchewan Wheat Pool Executive, 1982. Left to right: Garf Stevenson, second vice-president; Don Lockwood, first vice-president; and, Ted Turner, president.

secretary and the CEO. The president appointed members to the standing committees and chaired the board meetings. He was assisted in the role by a first vice-president and a second vice-president. Other than to open the meeting, the president did not chair the annual meeting of delegates. The president was the chief spokesman for the Saskatchewan Wheat Pool, gave major theme address to each annual meeting, and oversaw the preparation of the "Directors' Report to the Annual Meeting." He was extensively quoted in the media as the meeting ran its course, and was expected to answer any and every question raised in the ten-day annual meeting. He was aided in this by management personnel and also by the vice-presidents. The board of directors held ultimate power, but when the board was not in session, this power rested with the president or his designate (a vice-president, the corporate secretary, or the CEO).

The president/vice-president's offices, aided by the secretary's office, organized and apportioned duties concerning country meetings, governmental interaction, relations with other co-operatives and joint companies, as well as public policy organizations such as federations of agriculture and central co-operative assemblies. The president carried, as a matter of course, heavy responsibility for the entire co-operative; it was his choice and/or responsibility to delegate work. This system worked well for over 70 years because officials accepted responsibility for their levels and a willing staff did whatever was required of them.

Saskatchewan Wheat Pool Board of Directors, 1970. Ted, front and centre.

The board members worked hard to be informed and were not afraid to make difficult decisions. They also communicated effectively with delegates and the shareholders in their respective districts. One of the challenging jobs as chair of a board meeting is to keep board members from interfering with management. After some experience at the board level, directors recognized that on management matters, they could explore and comment, but they could not make decisions.

TRAVELS AT HOME AND ABROAD

The president was called upon to travel a good portion of the time, representing the organization, mainly within Canada, but often overseas, always remembering that his actions would be interpreted as those of the Saskatchewan Wheat Pool itself. The glamour of travel soon wore thin and a delayed flight would seem like an eternity. I was fortunate that for all my time in the position, Mel travelled with me as she was able.

The federations of agriculture—Saskatchewan and Canadian— were important conduits for advancing our public policy positions. I attended the annual meeting of the Canadian Federation of Agriculture (CFA) every year, and was one of the major debaters. It was here that I learned much about other parts of Canada and was able to forge links of respect and understanding. Every 18 months there was a meeting of International Federation of Agriculture Producers (IFAP). This took policy promotion to the international level. Here again, my understanding was enhanced and a fresh perspective from which to view our own situation was obtained.

I was a delegate to several IFAP meetings, including those in London, Paris, Vienna, Washington, Ottawa, and Adelaide. It was here that I first encountered the articulate delegates for the United Kingdom National Farmers' Union—Plum, Butler, and Winegarden. As a newcomer, I was unaware of their prowess or reputation and I challenged them vigorously about the European Economic Community (EEC) policies of subsidizing agriculture production and sales to the detriment of Canada. One of the delegates from the United Kingdom asked Wheat Pool director Aubrey Wood, "Where did you get that little bulldog?" Aubrey was quite proud of me. I came to understand, but not necessarily to accept, that western European countries, having been so desperately short of food during World War II, had vowed to decrease dependency on other parts of the world to supply their needs. The newly-formed EEC developed what was known as the Common Agriculture Policy (CAP) which, through subsidy incentives, encouraged food production even at a cost higher than could be purchased abroad. Europe welcomed surpluses and guarded producers against carrying costs; world grain prices declined because of the CAP.

I discovered that when the meeting got a bit dull, I could go to the microphone and attack the CAP, and before long, the Europeans were at the microphone, taking a strip off of me. This was similar to my occasional behaviour at CFA annual meetings. One day, the dairy producers had reported, and just before adjourning for lunch, I offered to give them canola oil to blend with butter fat into margarine. Even though I did so in jest, and such was recognized by many, others could not let it go until they set me straight; we were a half-hour late in adjourning!

These were magnificent meetings and I made many friends from other parts of Canada and internationally. I was proud, indeed, when Charlie Munro of Ontario, and some years later, Glen Flaten from Regina, became president of the IFAP, and prouder still of the outstanding job they did. While the policy statements flowing from the IFAP were profound, the real value of the sessions was the interaction among participating countries.

Another experience, although of lesser magnitude, was representing the CFA on the British North America Committee. This was comprised of 30 people from the United Kingdom and 30 from North America (10 represented Canada and 20 represented the United States). We would meet every year, alternating sides of the Atlantic. The main beneficiaries were the participants, as a result of frank exchanges among the countries represented. My last attendance was at Glen Eagles, near Perth, Scotland. Mel accompanied me and we experienced wonderful sight-seeing and golfing times together.

In 1996, at the encouragement of Allan Blakeney, I attended an agriculture conference at Ditchley House near Oxford, England. Kerry Hawkins of Cargill and I were the only Canadian participants. I was delighted to see that Lord Plum, a friend from the IFAP days was chair of the session. He was also an elected member of the European Parliament, headquartered in Brussels. Kerry Hawkins weighed in on the subsidy issue, which I noted was sensitive and brought forth an immediate vigorous response, just as it had a decade or more earlier. It was an interesting three days. Mel was able to accompany me and afterwards we enjoyed a few days in the Cotswolds of western England, and five days in London.

So, I was privileged to experience many overseas trips. As president of XCAN Grain, I visited London and our office there on many occasions to promote our exports or to iron out problems. On two such trips, along with the CEO of XCAN, John Hasselaar, we visited Europe to link with brokers in Germany and Holland. We formed several very productive partnerships. Also while I was president, we opened a sales office in Japan, where we set up Eddie Umemoto as our manager. Eddie spoke English very well and was highly respected in Japan.

Mel went along too, and witnessed the Japanese custom of the opening ceremony. The guests were received by Eddie, and the lower he bowed, and the longer he stayed down identified how important the person was. I am sure a few times he had carpet on his head when he straightened up! This office greatly enhanced our grain sales, especially of canola. It was interesting to meet Kim, Eddie's wife, and their small children. The office was a wise move and served XCAN very well.

On two occasions I visited China as part of a government of Canada delegation. In 1972 I went to the Solo Canada Trade Fair (Canada only) as part of the Canadian Wheat Board group. We easily had the most popular booth at the fair. There were some 400-plus Canadians in attendance. Our popularity stemmed from two facts. First, we had a miniature train running from a prairie elevator, through the mountains to Vancouver, which attracted an audience. We manned the display for 10–12 hours every day. Second, we were popular because Bill Coleman, a Canadian Wheat Board employee, could speak flawless Mandarin Chinese. This seemed to fascinate the audience. I spoke to many small groups through an interpreter, explaining the Canadian system, and had a slide show to best explain our system.

These meetings went very well. Twice we were called upon to attend formal meetings with Chinese grain officials. They were upset because some shipments were delayed due to labour disruptions. The Chinese wanted our assurance that such would not happen again. They would not admit that they understood how there could be labour disruptions. "We don't have any problems with labour" was their retort.

The Trade Fair was a great event, and the train trip back to Hong Kong was marvellous. The road bed was composed of concrete ties and the speed of the train was close to 100 miles per hour, with absolutely no vibration. Too soon I arrived in the Northern Territories of Hong Kong, and from there, on home.

In 1982, Senator Hazen Argue took a delegation of individuals involved in the grain trade to China. We visited Beijing for several official ceremonies and had many formal meetings with pointed questions about our reliability as a supplier— even though we had two decades of never missing a shipment. Chinese government officials were also concerned about weights, claiming that our shipments unloaded lighter than the amount for which we billed them. In a visit to a port, Forest Hetland of the Canadian Wheat Board pointed out to them that their hand unloading methods caused leakage that we witnessed.

It was also claimed that our wheat quality didn't always measure up. It was a common complaint which was never substantiated. It seemed that the Chinese always had some point to make.

We also officially visited Harbin and Shanghai. There was ample opportunity for sight-seeing and I once again visited the Great Wall, the Summer Palace, and other attractions. Mel was along, as were several other ladies, and outside of being challenged with some foods at the banquets, we enjoyed every minute of it. It was a great public relations trip on behalf of Canadian grain. Mel continued on to Singapore and then to Australia as I travelled home.

THE THREE POOL RELATIONSHIPS

The three Pools together undertook numerous functions and we had twice-a-year meetings where the three top-elected officials from each company met to deal with such things as Pool Insurance Company and Canadian Pool Agencies—companies we used to look after our insurance needs. We also met as Western Cooperative Wheat Producers to identify public policy needs and to strategize how to advance those needs.

I was quite compatible with the people from the Manitoba Pool Elevators (MPE) and the Alberta Wheat Pool (AWP). Often the three presidents—Harold Sneath of the MPE, Gordon Harrold of the APW, and me—would meet with industry or government officials. I was somewhat startled at the first meeting when it became clear to me that I was expected to take the lead role. This, in spite of the fact I was junior to the other two presidents in terms of experience. We each became very good friends. Harold retired after a few years and Gordon died while representing the AWP at a meeting in Banff.

Gordon and I had a remarkable relationship. He always deferred to me in meetings, but was wonderfully supportive. He had a fabulous memory, was able to think clearly and make sound decisions. He was, however, shy, and preferred to be one step back from the front line. Many of our meetings were in Winnipeg, so we often spent evenings together, having a sip of Scotch and rehashing the day's activities. As well, we were together at overseas meetings of the IFAP—often accompanied by Jessie and Mel—or in Geneva as advisors to the Canadian government on grain matters. Following Gordon's demise, I spoke to the annual meeting of the AWP. This was a common practise, and on this occasion, I spent considerable time eulogizing Gordon. Many delegates afterward told me it was the best speech they had ever heard and that they really appreciated the tribute to their former president.

FARM AFFAIRS IN SASKATCHEWAN

When I was elected president in June 1969, the effective date for the election was set as July 1, 1969. Having served as a vice-president for almost three years, I was familiar with office procedure, understood the scope of the president's role (but not entirely its complexity), had established positive relationships with key management people, and was favourably perceived by staff. Nonetheless, I got off to a somewhat nerve-wracking start.

Meeting with Prime Minister Pierre Trudeau, during his mid-July visit to Regina, would be my first official duty. Our publicity department/Communications head, Ian Bickle, had been able to arrange for a half-hour meeting for me with the prime minister. While nervous, since this was my first official duty as president, I soon relaxed and enjoyed the discussion. We had an excellent meeting in his suite of rooms at the Regina Inn. I found him attentive as I described the grim situation grain farmers in particular were facing. There were burdensome stocks of wheat worldwide, and thus markets were compressed and prices low. He asked good questions and encouraged us to work closely with the relevant cabinet ministers. I was impressed with his intellect and his demeanour. There was no easy solution to the problem, but it was important for the prime minister to hear the problems from a farmer's perspective. While I met with him, a demonstration occurred outside the hotel. In a loud and forceful manner, the National Farmers' Union (NFU) had organized people from around the province, and even beyond, to descend upon Regina to make known the need for action on farm problems. There were demeaning signs and insulting remarks—it was typical of the NFU's approach.

Following the meeting, many NFU members criticized me and the Saskatchewan Wheat Pool for meeting privately with the prime minister instead of joining with them in the protest. They were particularly upset about a remark Trudeau had made, "Why should I sell your wheat?" He went on to say, "you have the Canadian Wheat Board that I support and you have your own grain companies—the Pools." They chose to ignore his full statement and focussed on Trudeau's rhetorical question as a statement of indifference. Conflict with the NFU would haunt me all the days of my presidency. It was not a conflict of purpose, but rather a matter of method and style. We would have been derelict in our duty if we had not sat down with the prime minister when we had the chance to do so. The NFU felt that it could embarrass or force people to adopt a position, whereas the Saskatchewan Wheat Pool tried to achieve the goal by reason.

The friction between the Saskatchewan Wheat Pool and the NFU had existed long before I arrived on the scene. The NFU had brought Aaron Shapiro, often described as the father of the Wheat Pools, to the prairies to rally farmers into setting up their own grain handling co-operatives; thus, the three Wheat Pools. Once organized, the democratic structure that controlled the Pools was a natural conduit to recognize and advance public policy matters. This rankled the parent, the NFU, which saw farm policy as its exclusive domain and felt that the Pools should stick to grain marketing. This was a constant theme for them and when the Pools refused to use extreme tactics, they then became the targets of the NFU's wrath.

To compound the situation, NFU members were among the very best Pool patrons; the conflict all came down to style. Most farmers are reasonable people who first try to discuss, as witnessed by the make-up of the Saskatchewan Wheat Pool delegate body over the years. The Pool always contained more radical elements, but they were unable to enforce their extreme views as Wheat Pool policy. The NFU, over the years, had some remarkable leaders, who were capable of reasoned and effective debate, but unfortunately but they always chose the confrontational approach.

THE FARM CRISIS

The late 1960s saw the western farm economy plunged into crisis. Farm net income was almost non-existent, mainly due to low grain sales. In 1969, the worst year of the crisis, the Saskatchewan Wheat Pool had a net loss of $300,000 compared to earning $10,000,000 the previous year. Normally, the Country Elevator Division was the largest earner but it escaped the year 1969 with only a very small profit. Grain movement had to increase to rescue farmers and to restore the Saskatchewan Wheat Pool to profitability. The following year saw increased grain movement with a favourable increase in net income for farmers and net earnings for the Saskatchewan Wheat Pool at over $4,000,000.

This desperate period of low-volume grain sales and a build-up of world grain stocks brought action from the federal government, spearheaded by Otto Lang, minister in charge of the Canadian Wheat Board. The government initiated a study on Grain Handling and Transportation and created a Task Force on Agriculture, headed by Professor Clay Gilson at the University of Manitoba. The Saskatchewan Wheat Pool, which was always involved in all aspects of grain handling and transportation, critiqued the Task Force report. These two studies were to have a long-term effect on farmers and the grain industry.

A more immediate response was initiated by Otto Lang to deal with the burdensome wheat stock situation. The presidents of grain companies—both private and the three Pools—as well as the United Grain Growers and the Canadian Wheat Board, were invited to a meeting in Ottawa to be introduced to the Lower Inventory for Tomorrow (LIFT) program. This had been put together by Lang's staff, and in essence would encourage or force farmers to leave their wheat acreages idle for one year. Farmers would benefit from reduced costs and special delivery opportunities. At first glance, it seemed to make good business sense in that, by shorting the market, prices should rise. We were to meet each Saturday for the next few weeks to refine the program. This effort at "supply management" was surprising because many of those involved were usually critical of marketing boards for that very reason. Eventually a program was approved and put into effect for the 1970 growing season. At that time, the normal cropping rotation was either one-third summerfallow and two-thirds crop, or half-and-half. This approach then meant that considerable acreage would be summerfallowed two years in succession—not a desirable agronomic practise.

I took the lead, soon followed by others, that there must be a financial incentive. We finally settled on $6 per acre for each acre of summerfallow, and so it was put in place. The farmer reaction was swift and critical—they didn't like it! Over the next few months, all the company presidents who served as advisors to the minister, except for me, broke ranks and criticized the program. I was the lone exception for two reasons: one, I felt that inventory control was an effective practise. After all, it was used as a matter of course by manufacturers and in resource industries such as forestry, oil recovery, fisheries, and so on; two, I had given my word that I would support the program.

The final result was to ease the grain glut slightly, and it was not as devastating as farmers had feared. I was chastised by some Pool members but not severely. The major benefit was that government officials realized that the Saskatchewan Wheat Pool was reliable, for we had kept our word and our ethical behaviour was respected. The lesson here was always to be honest and not go back on promises: "People will forgive mistakes but not dishonesty."

In the meantime, the two major studies—the Grain Handling and Transportation study and the Task Force on Agriculture—would lead to long and protracted debates that pitted farmers and their representatives against government and some other factions in agriculture. There was no doubt the government wanted to solve the grain income problem, but it sought to do without incurring much public cost.

Those representing "prairie interests" were outnumbered in the federal cabinet, so they had to demonstrate that any proposed solutions would not have long-term cost implications and would not incite the call for similar money to go to other regions. Otto Lang, respected by Prime Minister Trudeau, was given the task of finding a solution. He was easy to work with as long as one accepted his approach. We had a good relationship coming out of the LIFT program, but this started to deteriorate when he began sniping at the Canadian Wheat Board. It further eroded when he sought to move transportation benefits from grain farmers to other sectors of agriculture by way of acreage payments. He took this stand, he claimed, to be production- and market-neutral.

Lang also sought to reduce the influence of the Saskatchewan Wheat Pool by inspiring the organization of the Palliser Wheat Growers, a product of the adherents to the Winnipeg Wheat (Commodities) Exchange. He established a personal advisory committee in Regina to offer him advice on policy matters for Saskatchewan. This committee, comprised of bright young lawyers, most of whom did not understand agriculture, would, when asked about the Saskatchewan Wheat Pool positions, respond: "They are anti-Liberal and that is why they take this position, or they do so to advance the Saskatchewan Wheat Pool." Nothing could have been further from the truth, and so the confrontation began. I will deal further with this confrontation when I explore fully the Crow's Nest Pass Rates.

CONSOLIDATION
In the 1968–71 period we had stagnated as far as increasing market share of country elevator grain handlings. All grain companies were experiencing low or non-existent profit levels. Under the leadership of Ira Mumford, CEO of the Saskatchewan Wheat Pool, we were engaged in activity that saw the number of elevators reduced from 1200 to 800, with plans to reduce further to 400 elevators. This approach had been widely discussed with the membership, and while they did not like to lose their local elevators, they recognized that under the present configuration, there were service limitations at small points.

Addressing a typical country meeting—
here, the opening of an elevator in Assiniboia in 1976.

I was party to many closing meetings and, without exception, there were harsh words thrown our way. One such meeting occurred in my own sub-district at Ruddell. This point had a very small area—Maymont seven miles to the east and Denholm seven miles to the west, the North Saskatchewan River four miles to the south, and another rail line five miles to the north—and we simply could not justify continuing to maintain this elevator. All other businesses had left and the school had been closed; only the Saskatchewan Wheat Pool elevator and the post office remained. (Often, we were the last business out of a community.) We were well-equipped with facts and a flip chart. The area superintendent, Herb Cobb, and I presented the bad news. The most vocal opponent was the postmaster, who was nearing retirement age and likely feared that if the elevator went, the post office would not be far behind. The chair of the local committee was very fair; he was a good operator in his own right, and expressed regret, but being a large farmer, also needed better service. Not so his cousin, a volatile person, who rose and criticized me, saying I was only doing this because it would benefit my home point of Maymont. There was dead silence. My only response was, "I am sorry you feel that way, Willie." The meeting carried on, and a few minutes later Willie got back on his feet. He said, 'I am sorry for what I said, Ted. You have always served us well and what I said was not true. I am sorry." Needless to say, the meeting was over and the community reluctantly accepted the closure.

We continued to explore ways to improve our grain handling percentage and also our return on investment, and it was suggested that we purchase one of the private companies. We knew that they too were under financial pressures because of lower-than-desired volumes. We targeted Federal Grain Ltd. It was the largest of all the private companies, covered the three prairie provinces and would give us the greatest leverage in sensibly rationalizing our handling network. Like the Pool, Federal had 1200 country elevators but they were spread across three provinces. They also had good export terminals at Thunder Bay and Vancouver. They were purported to be the best managed of all the non co-op companies, and we knew they were terribly dissatisfied with their return on investment, so it was worth an attempt to buy them out.

FEDERAL GRAIN PURCHASE

Grain handlings were controlled by the amount of grain produced, the amount that had been sold domestically and into international markets, or by a combination of those two factors. Operating costs were rising and handling charges were controlled by the Canadian Grain Commission, which was unlikely to grant significant increases. The only offset was to increase the volume of grain handled. Grain car shortages created a sort of rationing between companies. This generally worked to our detriment; Saskatchewan Wheat Pool elevators, at a shipping point, would fill up first and then the opposition companies would benefit from our inability to service our members because of our lack of space. So, at the end of each year, our market share was about the same as the previous year. Adding storage space only helped until the new storage space was filled. Farmers were fiercely loyal to their home towns, so it was very difficult to close a "home point" and to expect that our members would follow to an adjoining point. There were facilities traded between companies and this type of elevator consolidation helped, but not nearly enough to offset rising costs. Something had to be done, or the system would crumble before our eyes. Those government officials calling for massive consolidation, as were the railways, simply did not understand producer loyalty to their home points.

This whole situation of being able to meet cost increases prompted Ira Mumford, to write a memo to his counterparts in the Alberta and Manitoba Pools, suggesting that perhaps we should attempt to purchase the largest publicly owned elevator company, which was Federal

Grain Limited. Mumford noted that chronic elevator congestion had almost eliminated competition between companies. With the decline in profits, drastic action was called for if the Pools were to maintain charges at a reasonable level for producers and to provide an upgraded level of service. He advised that the Pools act quickly, before we were pre-empted by one of our competitors.

A memo to the Saskatchewan Wheat Pool Board of Directors identified that there were 217 delivery points in Saskatchewan, where the acquisition of Federal Grain Limited would give the Saskatchewan Wheat Pool a monopoly situation. Further, Federal handled 16 percent of Saskatchewan grain; the memo suggested that the Saskatchewan Wheat Pool, because of congestion, would retain all of that grain. Each of the three Pools reacted favourably to the suggestion. Now, the question became: "Who was going to bell the cat by approaching Federal Grain Limited?"

At a meeting of the executive of the three Pools, George Turner (no relation), president of Manitoba Pool Elevators, related that there was talk that Federal Grain might be for sale. After a brief discussion, George was asked to contact George Sellers, president of Federal Grain, to ask him to join our meeting. Later that afternoon, Sellers, as he described it, "slunk" into the Royal Bank Building that contained the head office of Manitoba Pool Elevators. Our three Pools told him of our interest in purchasing Federal Grain. He said that the return on their investment was unsatisfactory and he would talk to his colleagues and get back to us.

Sellers got the blessing of Federal Grain officials and talks began in mid-1971, only to be stopped because of disagreement on certain issues. Then, in early September, I was awaiting a flight to Regina from Winnipeg when George Sellers disembarked from an arriving plane. After a few pleasantries, I asked him if it was time to start negotiating again and he said, "Yes, it is." My first action, once back at home, was to phone Ira Mumford and to relay my contact with Sellers. Ira contacted Wally Madill, CEO of the Alberta Wheat Pool, and Bob Moffat, CEO of Manitoba Pool Elevators; the negotiations started again.

Ira Mumford was the key individual in directing the negotiations, and he would consult either by telephone or in person with a negotiating committee that was comprised of the president and CEO of each of the Pools. Ira dispatched his executive assistant, Milt Fair, and the

Saskatchewan Wheat Pool corporate lawyer, Bob Milliken, to Winnipeg to do the detailed negotiation. They did a magnificent job, and in the end, there were no loose ends to clear up. During the annual meeting of delegates in 1971, I spent considerable time in a Hotel Saskatchewan room talking about the negotiations and giving my input to the negotiating team. Negotiations were at a critical stage and the fact that our annual meeting and the meeting of the Alberta Wheat Pool were each underway made it difficult for Ira to assemble his "negotiation committee." Following the adjournment of our meeting one day, Ira, Milt and I flew to Calgary, where we met with the full negotiating committee until midnight. We flew home early the next day and were present when our annual meeting reconvened, and no one even knew we had been gone.

Because Federal shares were publicly traded, secrecy of the negotiations was absolutely critical. Our annual meeting was adjourned on Friday, and the next morning Ira and I flew to Winnipeg where, after a day-long meeting, we essentially concluded the agreement to purchase Federal Grain. However, there was still much to be done in covering all the details and in committing words to paper. It became increasingly challenging to maintain secrecy as time wore on. We deliberately under-informed our board on the status of negotiations in order to protect them from being caught in a situation where they had to deny there were negotiations.

Gordon South, a Saskatchewan Wheat Pool director, was asked prior to the purchase announcement about a rumour that such was happening. Gordon reported later that he had to lie, probably for the first time in his life. Finalizing the documents continued with more than one crisis point, but all were successfully handled, and by mid-March, 1972, everything was in order.

It was a tremendous thrill on March 16, 1972, to sign two cheques— one for $90 million that included all inventories—and one for $30 million that was to cover the cost of the negotiated deal. I returned home later that day with a bouquet of flowers in hand, to celebrate Mel's 45th birthday.

The deal was very comprehensive: (1) Federal employees were treated fairly and many of them eagerly became Saskatchewan Wheat Pool employees; (2) Federal Grain's chief engineer, Michael Thompson, became the Saskatchewan Wheat Pool's chief engineer, and later would become CEO of the newly constructed Prince Rupert Grain Terminal; (3) we assumed all outstanding obligations to Federal Grain employees and transferred pensions where applicable; (4) the Saskatchewan Wheat Pool purchased Federal Grain's accounts receivable for fifty cents on the dollar—mostly farm supplies—and made a good profit on the transaction; and, (5) the Pools got confirmation that our management structure and operations were superior to that of Federal Grain's, as demonstrated by the fact that our negotiators were in-house people —Milt Fair and Bob Milliken— while Federal Grain used its corporate counsel and the head of its audit firm. Federal was regarded as the best-managed of all the private grain companies. Ira Mumford later reported that the Saskatchewan Wheat Pool probably recovered its investment of $15 million (the total deal was for $29.5 million) in the first two years.

To be sure, there was fall-out from the purchase. I received a barrage of letters complaining about the fact members and even delegates were unaware of the action and/or that competition would be reduced. I answered each letter within a day and was never taken up on my offer for a face-to-face discussion of the matter. People relaxed when there was no precipitous action to close elevator points. Most of our members felt a surge of pride from the fact that their co-op was now recognizably the largest grain company in Canada, and perhaps the largest primary grain company in the world.

The purchase met our objective of retaining all of the Federal grain business as our handling percentage in Saskatchewan rose from 50 percent to 66 percent. Of course, other elevator companies expressed dismay and unsuccessfully tried to lure our members to competitive points. Some newspapers, by editorial comment, condemned the deal. I was disappointed that the Regina Chamber of Commerce did not laud the fact that with a stroke of a pen, the Saskatchewan Wheat Pool had transferred a $100 million business from Winnipeg to Regina.

The seeming discontent in the country soon dissipated as it became apparent that we were not going to suddenly close a number of elevator points. Increased patronage dividends were balm for the wounds of many disenchanted members.

The purchase of Federal Grain increased our numbers but most importantly gave us the ability to reduce outlets and still give upgraded service to local points. The 400 points, while a physical target, did not have a definitive time line. We were quite prepared to let the system evolve with our guideline being adequate service to members consistent with sound financial management. This would be knocked into a cocked hat years later, when the railways were allowed to use variable freight rates and could thus dictate the configuration of the grain-handling collection system.

PERSONNEL AND RECOGNITION

During the early 1970s my secretary, Rose May, retired and I was sure I would not find an adequate replacement. However, the search process identified Kay Klotz as the most worthy of four very capable candidates, and thus started a relationship that transcended the office. She became a family friend and also a close friend to Vera, Mel's Australian pen pal. Kay and Vera were each devout Catholics, so they had much in common. Their pleasant personalities blended perfectly and the friendship lasted until Kay's death in 1996 at the age of 69.

Kay fully understood her job and mine. She seemed to know when to protect me from over-involvement, and who it was important for me to see. She was technically very competent, fast and accurate, and completely loyal. On occasion, she would come into my office with a letter that was abusive to me in her hand and tears of anger in her eyes. She would say, "How can he be so vicious in his remarks? He doesn't even know you!" I would take the letter from her, and depending on my mood and work demands, would either respond then or in three days. We had a policy that, unless I was away, every letter would be replied to, or at least acknowledged, within three working days. The more abusive the letter, the softer the reply, without compromising anything and mostly ignoring the worst language. Kay was a gem and I was lucky to work and to be associated with her for 15 years.

Upon receipt of a particularly "bad" letter, I would, on occasion, arrange to meet personally with the letter-writer by making an appointment at the person's farm. These individuals were usually quite different face-to-face than they sounded in the letters. Each time I left with a "thank you for coming; I am sorry I was misinformed or didn't understand."

One frequent correspondent was Jack Stueck of Abernethy. He would write any time something happened that caused him concern. His language was strong, his points not well taken, and his letters ran to five or more pages. Rather than fret about his remarks, I treated his

epistles as a conversation and replied right away, ignoring his most vit-
riolic comments, and thanking him for his letter. He said, in one letter,
"Why do you reply so quickly? It gives me less time to be angry!" I did
meet him personally and found an individual who, while explosive and
vocal, was concerned about others and would host young farmers from
other countries to help them understand the Canadian way. I think, but
am not sure, that we became friends.

My office life was very nice. I established good relationships with all
the people with whom I had direct work contacts. When I met others,
I would smile and greet them. There was a deep respect for authority
in our environment, so I always tried to put people at ease and, I think,
was successful most of the time. Over the years, I was able to call most
of our employees by their first names.

Each year, we held a series of long service awards that recognized,
by way of a lovely watch, 25 years of service. For 30, 35, and 40 years,
there were lapel pins; there may have been small monetary awards as
well. These would take place for Regina employees at the time of our
September board meeting. The board, each year in late September or
October, would make a trip to Thunder Bay to inspect facilities. There
would be a large "service awards" banquet with community dignitaries.
Then, on the way back to Regina, a similar event would be held in Winnipeg. Later, we would do the same in Saskatoon and Vancouver.

These events were well done; all employees with 25 years' service
were invited to the banquet in their area. We tried to make them all feel
special. Often in Regina and Thunder Bay, there would be as many as 30
award recipients. It was a very heavy night for the president, who acted
as master of ceremonies, made a short address, and also presented the
awards. After a couple of years, we changed that procedure so that one
of the vice-presidents would chair, the other vice-president would make
the address, and the president would make the award presentations.
Although a lot of work and expense, these events generated a great deal
of good will, and created a better understanding of how very important
our employees were to us.

KEY RELATIONSHIPS

During my years at the Saskatchewan Wheat Pool, I was privileged to
work with many very fine people. Charlie Gibbings was president during my first nine years as a director, which included three years as first

vice-president. He was a good friend and an excellent role model. The delegates, especially in District 16, over my span of 29 years, were for the most part inspirational and some of the friendships last to this day. They were, on balance, wonderfully supportive.

The numerous directors with whom I shared the board table were truly dedicated to the development of the Saskatchewan Wheat Pool for the benefit of farmers. The vice-presidents with whom I worked—Ted Boden, Don Lockwood, Bill Marshall, Garf Stevenson, and Avery Sahl—were capable individuals and we had compatible relationships.

Senior staff—Ira Mumford, Jim Wright, and Milt Fair—left nothing to be desired, were always helpful, were blessed with good judgement, and wanted only what was best for the Saskatchewan Wheat Pool. Rose May, Kay Klotz, and Sonia Korpus, my personal secretaries, could not have been more loyal and competent.

I had a warm and inspirational relationship with all the division managers, especially so with Ian Bickle, Allan McLeod, Duane Bristow, Chris Hansen, Ron Kasha, Bob Phillips, Jim McDonald, Ron Sproule, Ken Sarsons, David Wartman, and Glen McGlaughlin. Also, at the division level, Mel and I became close friends with Charlie (Doris) Leask and Roy (Dorothy) McKenzie. Other senior level support people, John Trew, Don Sinclair, Marj Dickin, Don Ross, Bonnie Dupont, Connie Gervais, Jim MacDonald, Ron Sproule, Scottie Graham, Harold White, Mac Bjarnason, and Hank Brown were outstanding. I also worked closely with, and had great support and inspiration from Lorne Harasen, Marylin Reddy, Metro Kereluke, Ken Orr, Doug Kirk, Duane Mohn, Bryce Belt, John Derbowka, Warren Crossman, Al Laughland, Ron Weik, Howie Hatton, Nial Kuyek, Glen Peardon, Murray Bryck, Martin Hopkins, Ian Traquair, Mac Lambie, Clare Pyett, Bert Lee, and Ivan MacDonald.

There are others equally dedicated and capable who are too numerous for individual mention at this time. I respected all who worked for the Saskatchewan Wheat Pool and all who were part of our democratic structure.

CONCLUSION

As discussed above, I had the good fortune during my tenure as an elected member of Saskatchewan Wheat Pool to work with many wonderfully competent people, both within and outside of the company.

Sometimes, inevitably however, advancing the interests of Pool members brought me into conflict with others: perhaps elected or appointed government officials, individuals from competing companies, others from companies in the service industry (such as railways), sometimes journalists, and sometimes even people from other countries. It would have been easy to conclude that they were all against farmers, or opposed to cooperatives or just wanted to see SWP fail, but in reality they were doing exactly what I was doing—that is, they were looking out for the best interests of those they represented.

In such cases I tried to look beyond the spoken words and focus on the speaker. Most often I found sincere and dedicated people who were simply doing their job and only from a biased perspective did they appear mean-spirited or devious. It was my challenge to assist them to understand our point of view.

Throughout my career I witnessed, and indeed participated in, many dramatic changes within the agriculture industry and in the functioning of the agricultural economy in Canada. But perhaps the longest running and most impactful change at play during my career was the broad issue of grain transportation. The initial grain collection/shipping configuration had been put in place at a time when farmers hauled their grain from their farms to elevators using horses or oxen. Thus, rail lines and elevators were built very short distances apart. The complex network that evolved was the result of keen competition between railways and equally keen competition between elevator companies, both of whom attempted to woo farmers with shorter travel distances. This action resulted in an over-built system that denied elevator companies the rewards that come from economies of scale since the close proximity of one elevator to another prevented a significant level of business at any one shipping point. Their ability to load a large number of cars was curtailed by low stocks or improper grade. The railways on their part found it difficult to maintain a high mileage of track and a large number of sidings; it was especially costly on low volume branch lines where grain was perhaps the only commodity hauled. In the post-World War II period there was a marked increase in international demand for Canadian grain which in turn required a corresponding increase in grain movement to port positions. Our extravagant system that was spurred by competition became an impediment to efficient grain movement. Elevator companies responded by dramatically reducing the number of shipping points and by building fewer but more efficient plants. The railways moved to abandon branch lines and thus reduce the time required to collect the needed grain and assemble their trains. This foreshadowed the use of efficient unit trains that were the ultimate in grain movement. (A unit train is a train in which all the cars making it up are shipped from the same origin to the same destination, without being split up or stored en route.)

In each case there was strong resistance from the farmers most effected, who felt that they alone were carrying the burden of increased grain movement efficiencies. This created a dilemma for farmer-owned cooperatives because members felt we should protect their specific interests rather than make judgements about the greater good. They were unable to separate the concept of fair treatment from the idea of treating every member the same, nor could they accept that equitable treatment did not mean equal treatment. I had many heart wrenching moments from such situations.

In 1957, my year of entry into matters "beyond the farm gate," Thunder Bay was the dominant export terminal for prairie produced grain. The sale of large quantities of wheat to China in 1962 marked the start of a shift in exports from the St. Lawrence Seaway to the pacific seaboard. It would be many years before the full impact of this trend was felt; however, the dynamics of grain movement were altering how resources would be deployed and dictated that major changes were required in all the basic components of moving grain from a farm to a ship. Thunder Bay would no longer be the dominant export terminal and its huge storage capacity would no longer be required. Companies had to face new challenges and struggled to put into place the necessary infrastructure to maximize grain shipments via Pacific coast ports. The corporate interests of Saskatchewan Wheat Pool were always best served by a free-flowing movement of grain from country elevators to export terminals, and earnings would be increased by handling larger volumes. Additionally, the lack of congestion at country elevators allowed SWP to gain market share.

However, the free flow of grain could potentially be hampered by several factors: decreased international demand for Canadian grains; the ability of export terminals to have specific grain products available for shipment; railways lacking sufficient rolling stock (both grain cars and motive power); a collection system for grain that made it difficult to move desired grains and quantities expeditiously; strikes by long-shoremen or grain workers; and, severe winter weather conditions on the prairies and in the Rocky Mountains (extreme cold and snowfall reduces railway operating efficiencies).

These difficulties had the potential to strain membership relations within SWP as members were best served by a free-flowing movement of grain from Saskatchewan to port positions. Farmers became cranky when grain movement was slow and even quite abusive to their own officials if they perceived that actions or inactions by SWP caused the holdup. It was a member's right to complain, as it was their bottom line or even survival that was at stake.

Lack of grain movement created economic problems for all communities on the prairies. Therefore, provincial governments could not resist, nor should they, the temptation to intervene on behalf of their citizens. During one period of slow movement in the latter part of the 1970s I was invited (an invitation I could not refuse) to a special meeting called by Premier Peter Lougheed of Alberta, Premier Allan Blakeney of Saskatchewan, and Premier Sterling Lyon of Manitoba to discuss grain transportation problems. Also attending were Alberta Wheat Pool President Alan Macpherson, Manitoba Pool Elevators President Wallace Fraser, and United Grain Growers' President Mac Runciman. In addition, officials from the Canadian Wheat Board and the Canadian Grain Commission also were present. Since the meeting was in Winnipeg, Sterling Lyon was chair. The premiers were very fair, and it appeared they learned a lot about the complexities of the situation. They indicated a desire to assist in any way possible, and, as such, it was a very positive meeting. The session brought increased attention to a chronic problem, and it served to soften the attitude of taxpayers to possible expenditures by government, both provincial and federal, toward helping to resolve the situation of inadequate grain movement.

All sectors in the grain industry participated in building a lasting infrastructure that would serve the need to move prairie-produced grain in an efficient manner to west coast ports as well as its expeditious disbursement onto ocean-going vessels. The Governments of Alberta and Saskatchewan purchased grain hopper cars and contributed them to the railways' collection fleets; additionally, the Canadian Wheat Board purchased hopper cars and put them, too, at the disposal of the railways. Both CN and CP also leased additional motive power, and

CP completed the Rogers Pass Tunnel that saved both time and money in moving grain to Vancouver. Grain handling companies increased storage and shipping capacities at west coast ports, SWP and Pioneer Grain each built a new terminal, while Alberta Pool and Pacific Elevators upgraded their plants and a consortium built a large state-of-the-art terminal at Prince Rupert. In turn, farmers accepted, although often reluctantly, longer distance hauling to elevator collection points. All the while, senior railway officials made bold statements in public forums claiming that if the statutory rate for transporting grain was replaced with a remunerative rate their companies would never again be found deficient in the movement of grain. In 1995, after years of debate, farmers finally endured the demise of the "Crow Rate," albeit with the hope of achieving a system that would better satisfy their needs.

As new grains were added to the mix of commodities to be sent to export, so, too, there was added complexity to operating the system. While it seemed that too often there were periods of concern, the system expanded its capacity never-the-less. No doubt grain movement from the prairies to export positions will always be a 'work in progress'—one that demands attention, risk, innovation, and investment.

CHAPTER 6

Saskatchewan Wheat Pool
Affiliations and Relationships

The three prairie Wheat Pools had much in common. They had all been created within months of each other, and all served the same client base. Initially operating as grain handlers, they returned a portion of each year's profits to the farmers who patronised them. Given their similarities, it was not surprising that over time the Pools undertook a number of commercial ventures together. There were areas where very successful joint ventures paid handsome dividends to the Pools based on their level of patronage.

WESTERN CO-OPERATIVE FERTILIZER LTD.

The Saskatchewan Wheat Pool's entrance into the farm supply business created the need to source product. It was obvious from the start that weed spray and fertilizer would be the two areas of greatest sales volume. Weed spray was available through a facility owned by Inter-provincial Co-operatives, which welcomed the additional volume that would be generated, but there was no such convenient co-op supplier for fertilizer. Each of the three Pools was entering the farm supply business at the same time. Federated Co-operative Limited (FCL) was already well-established in supplying the farm-related needs of local co-ops.

It was a major owner and the largest customer of Interprovincial Co-op's chemical plant. FCL also had a fertilizer supply arrangement with Cominco, a large fertilizer manufacturer. The Saskatchewan Wheat Pool was able, in the first few years, to hitchhike onto the Cominco arrangements. Then, suddenly, Cominco became hard to deal with. Supply volume and price became uncertain and placed the Pool in a non-competitive position on retail sales.

This could not be tolerated and led to serious talks between FCL and the three Pools about alternative sources of fertilizer procurement. The decision was taken to build a plant to manufacture enough product for the combined needs of the group. The plant came into production in 1965. After the governance structure was worked out and became functional, the joint venture (Western Co-operative Fertilizers Limited) proved to be very successful and served the needs of Western Canadian farmers for several decades. I had the privilege of serving on the board from 1969 to 1986. Besides the business at hand, Western Co-operative Fertilizers Limited also provided a forum for the Pools and FCL to meet.

XCAN GRAIN LTD.

For grains other than those handled by the Canadian Wheat Board, the prairie Pools had to rely on private traders for export. By the late 1970s, canola was becoming increasingly important as an export crop, and the three Pools handled almost 50 percent of Canadian canola production. At a meeting of the executive of the three Pools, Bill Parker, president of the Manitoba Pool, who was closely connected to the grain trade in Winnipeg, suggested that it only made sense to export our own product. He proposed that we buy a company called Ken Powell Ltd., a respected and successful Winnipeg firm with an operating office in London called Powell-Union. We were able to do so and formed XCAN Grain Ltd. Ken Powell stayed on for several years as the general manager. When he retired, Ed Pierce was hired as CEO, and the London office was expanded.

Pierce did a good job and put in place competent staff people. One such individual was John Hasselaar, acquired from Bunge Corporation, one of the world's largest grain trading companies. Hasselaar replaced Pierce when the latter retired in 1974. By this time, I was president of XCAN and it was my job to help John work smoothly with the three

Pools. John expanded the business dramatically. He hired a sales manager, Gus Deslauriers, who was very competent, and added Eddie Umemota, who would become the manager of our Tokyo office. Eddie had a great personality and was a skilled trader.

I went with John on several trips to Europe to meet potentially new trading partners or to help resolve one problem or another. John was not a polished businessman and he relied on me to handle the protocol of our situations. I grew quite comfortable with John and we were a good team—he had great trading knowledge and I was able to handle the rest of what was required, particularly the interactions with others. I felt that above all else, John was completely honest. Sadly, as the events of early October 1985 were to prove, I was mistaken.

A day or two preceding the annual meeting of XCAN Grain, to be held in Vancouver on October 2, 1985, I received a telephone call from John Hasselaar asking if it were possible for me to meet with him the evening of October 1 in Vancouver. We arranged for John to pick me and Milt Fair up at the airport upon our arrival at Vancouver, and take us to the Holiday Inn on West Hastings.

Following check-in, John and I met, and he asked if the auditor, Brent McLean, of Price Waterhouse in Winnipeg, could join us. I agreed and in the following discussion, John indicated that he had an outstanding loan of approximately $165,000 with XCAN (the exact amount was $167,360). I reminded John that there was no authorization for him to issue a loan to himself. He did not argue, but suggested there was a minute of a previous board meeting that indicated we might consider this, providing there was no significant administration cost to XCAN Grain, and that there was a satisfactory rate of return on the loan. The minute to which he referred dated from April 16, 1981. It clearly illustrated that the matter was discussed, but that the conversation had been closed by my suggestion that management should prepare a policy to be reviewed and approved by the board.

The two men then asked if I could ignore the unauthorized loan, since John intended to repay it over the course of the next few months. John informed me that he had signed a promissory note on a call basis filed with the company, that the interest rate was the bank prime rate, which in his estimation would provide a better return to XCAN than other loan monies.

I told John that if he did not report this to the meeting, then I would have to do so. The next morning John informed a stunned board of directors of the situation. Unfortunately, this would prove to be only the tip of the iceberg. I was startled when the treasurer stated that he also had a loan, as did several other individuals. We were left with the impression that this was a recent development, and that all loans would be repaid within a short while.

The discussion ended there, without a resolution, because the rest of the business of the annual meeting still had to be dealt with. The next day was the annual meeting of Prince Rupert Grain, and since many of the XCAN Board would be attending it, we felt we could discuss the situation in the intervening hours.

On Thursday, October 3, I placed a call to Price Waterhouse and demanded that their managing partner, Bob Plaxton, meet with Milt Fair and me at my office in Regina the following day. Plaxton was unavailable, but Jeff Lecuyer and Brent McLean showed up. Milt was as angry as I had ever seen him. He angrily said to McLean, "How dare you ask Mr. Turner to overlook the unauthorized loans?" McLean responded, "I knew he would not go along with the request." By the end of our meeting, they had agreed to do whatever was required to resolve the situation. They even produced a letter written a year earlier, advising John that this action was not authorized and needed to be resolved. In fact, after the letter, things got worse. McLean should have blown the whistle at that time, rather than to let it continue, but he believed John would repay his loan and cease authorizing other loans. Clearly, McLean was derelict in his duties.

There were, it turned out, loans to eight or more employees, not all of them secured, and some of Hasselaar's loans went back four years without any security, and had actually grown from year to year. The indication was that John had outstanding loans of $165,000 with a promissory note that covered $80,000. Further investigation revealed that travel advances and personal air flights for John and his wife raised the total to over $210,000.

Milt and I spent most of the weekend in the office, and after consulting with officials from the Manitoba and Alberta Pools, we agreed to hold an emergency board meeting in Winnipeg on Monday, October 7. The outcome was agreement by the board that John Hasselaar be fired. The board appointed Ken Sarsons, CEO of CSP Foods Ltd., as the acting CEO of XCAN Grain. Sarsons had a long experience with the Saskatchewan Wheat Pool, having come to the Pool as manager of the flour mill. When the vegetable oil plant and the flour mill were merged into the Industrial Division, Sarsons was named as the general manager of the new entity. He was a remarkably competent administrator with the ability to develop staff and at the same time grow the business. Within a few days I fired Hasselaar and introduced Sarsons to the staff as the new CEO. The tension around the office was unbelievable, with the CEO being relieved of his duties and the resignation of the treasurer. Both Sarsons and I addressed the assembled staff but it was doubtful that they heard very much of what was said. Two weeks later, along with Alan Macpherson, president of Alberta Wheat Pool, we returned to the office, told the staff that XCAN was important to the Pools, that it was a viable company, and we asked for their help in making the company even stronger than it was. There was a noticeable difference in attitude, as everyone seemed more relaxed. It was now possible to have one-on-one conversations. Obviously Sarsons was having a positive effect on the employees.

During the SWP board meeting on October 15, I received a phone call from the head officer of F-Division of the RCMP, asking if we could meet for lunch the next day. Joining us for lunch were Sergeant Earl Basse and another officer from Winnipeg. They took me into a room where a table 12 feet long and 4 feet wide was literally covered with files, all pertaining to XCAN Grain. What followed made the unauthorized loans seem almost insignificant.

Sergeant Basse went to a blackboard, and with the help of some diagrams, explained "circle trading" to me. A corporation (XCAN) would sell canola to another firm for, say, $100 per tonne. Firm B would then sell to Firm C for $100 per tonne. Firm C would sell to Firm A (XCAN) for $110 per tonne; thus, XCAN would lose $10 per tonne. The profit by Firm C was then divided by the people involved at the expense of XCAN.

When one considers most deals were for 1,000 tonnes, if done often enough, the figures were impressive. Basse asked me why XCAN was doing this, and if the actions had been authorized by the board. I replied that I had no idea why this was being done, and assured the officers that the board had not authorized them. I was shown over 100 files detailing similar transactions.

When I instructed the RCMP to get to the bottom of the situation they were delighted. When I asked why, they said in most similar cases the president would say "We will handle this internally. We don't want adverse publicity." Earl Basse and his colleagues were anxious to follow through. He also told me that they had agonized over whether to contact me because I could have been part of the scheme. When they learned that I had fired Hasselaar they felt confident that was not the case. I then phoned Ken Sarsons in Winnipeg and told him to catch the earliest possible flight to Regina. He arrived at 5:30 p.m.

Over the course of the next few hours Ken, the RCMP officers and I strategized our next move. Sarsons was superb and, along with the police, later delved into XCAN files in both Winnipeg and Vancouver.

The RCMP investigation eventually revealed that XCAN had been the victim of fraud for a staggering $3.1 million—a huge amount of money for a relatively small company to lose. John Hasselaar was arrested, found guilty and served time. While in jail he wrote me several letters expressing his regret and asking that I, on his behalf, apologize to the board of directors. On advice of counsel I did not respond to his correspondence. Following his release John had difficulty finding a job and that exacerbated the family and health problems he was having. Gus Deslauriers, director of trading, was fired and replaced by Peter Lloyd who did an even better job than Gus had done.

The RCMP were relentless in their investigation. After I had resigned from the Wheat Pool, as well as from Prairie Pools Inc (see chapter 10), I received a phone call from a Corporal Tario. He brought me up to date on the XCAN file and advised me that they wished to cast the net even wider. The RCMP were convinced that there were people other than John Hasselaar who should be indicted for the fraud of XCAN Grain. Tario indicated that they could not get Hasselaar to cooperate in the investigation and to give them the names of others who were potentially involved. He felt sure John would talk to me and so suggested that I arrange a meeting with him and that the RCMP record our conversation. At first I was reluctant to do so because it seemed so deceitful; however, I rationalized that John had deceived the farmers who were the owners of XCAN and that he had been willing to hang me out to dry on the unauthorized loans. John eagerly accepted my invitation and we met in a hotel room in Richmond, BC, that was miked, and our conversation was recorded in an adjoining room where Corporal Tario and a Constable Edwards were listening. Earlier, they had given me a list of questions that I was able to work into our discussion. John described his plight: he had been unable to find a decent job because of his time in jail; he had trouble walking because of an aneurism in each leg; and, there were family problems as well. It was an emotional meeting. John broke into tears several times as he proclaimed his remorse for his actions while CEO of XCAN. He pleaded for compassion from the XCAN board. I told him then as I had previously, that it was not possible for me to help him. I told him that his best course was to confide in the RCMP and to help them in their investigation. He was quite open in admitting that he had lost control of the company and was unable to handle Deslauriers who had on one occasion told him to keep out of the marketing area—this was subordination that should have got Deslauriers fired. John steadfastly claimed he had no knowledge of the whereabouts of the $3 million, saying if he had he would hire a really good lawyer instead of relying on legal aid. As we concluded our meeting my heart was heavy even though I had got him to reveal the names of some of the culprits. My parting words were, "John please contact the RCMP and cooperate with them. You should not be protecting people who, if they can, will shift all the blame onto you." John was noncommittal, but he sincerely thanked me for allowing him to "get [things] off his chest." After the meeting, I was startled when I went next door and saw two services pistols on the bed ready for use. The next day Tario and Edwards went to John's home and got much of the information they were seeking. Some time later there was a trial, but I do not know the eventual outcome nor do I know if XCAN Grain Ltd. ever recovered any of the $3.1 million they lost due to the fraudulent activities committed against them.

Our competitors were convinced that XCAN's name was tarnished and they tried very hard to take advantage of the situation, but to no avail. Contacts with our major customers relieved their anxiety and assured them of a continuing positive relationship. Under Sarsons' management and Lloyd's skilful marketing, the company functioned very well. The morale of staff was good, and contacts we made with customers in Winnipeg and through our offices in London and Tokyo resulted in keeping our market share and, indeed, positioned us to move onward and upward, and such was the case.

POOL INSURANCE COMPANY

Wooden elevator structures and grain dust created a fire hazard and each year an elevator company could expect to lose facilities to fire.

Insurance costs were high and rising, so Pool Insurance was established, to which the three prairie Pools paid the industry average rates for coverage. At year-end, very good surplus earnings were paid out on a pro-rata basis.

These benefits were further enhanced by the appointment by Pool Insurance of fire inspectors who, annually, would visit each country elevator facility and also inspect the large export grain terminals. This raised the consciousness of the facility operator about fire prevention, and there was a noticeable drop in the number of fires. This, however, did not lower the rate of convenience fires, a term used to describe the situation where an elevator agent would torch the elevator as a way to cover up a dishonest activity. The first action by management following a fire was to check the records for that facility. Never did such action go unpunished. Obviously, insuring close to 3,000 elevators spread the risk, but also created a vulnerability if, for some reason, there were many fires in one year. Also, if a large grain terminal was destroyed, the replacement costs would be in the millions of dollars. To provide a safeguard, Pool Insurance established Pool Agencies.

POOL AGENCIES

Pool Agencies operated as a broker and reinsured all of the facilities covered by Pool Insurance. Because of the volume of business, favourable reinsurance rates were available. Lloyds of London carried the majority of this secondary placement of insurance. Pool Agencies charged Pool Insurance average industry rates and returned a patronage dividend at year-end that became part of Pool Insurance Company's earnings that were, in turn, paid as a patronage dividend to each Pool. Over the years, these two companies returned millions of dollars to their members, so the Pools benefited from lower net costs of insurance and still received the desired level of protection.

PACIFIC ELEVATORS AND WESTERN POOL TERMINALS

Following the purchase of Federal Grain in 1972, it was necessary to deal with the assets acquired in that transaction. Federal had a sizeable grain terminal in Vancouver that was now the property of the three Pools. Apportioned ownership was worked out based on the expected usage of the new export facility. The original name, Pacific Elevator Ltd., was maintained. As the potential largest user of the plant, Alberta

Pool was asked to manage it. This was accomplished when Alberta Pool incorporated Pacific Elevator Ltd. into the management of its own terminal. The President of Alberta Pool was always the President of Pacific Elevator Ltd.

The Saskatchewan Wheat Pool CEO and I were on the board, as were the president and CEO of Manitoba Pool. Cargill Grain, with which Federal had a handling agreement, was also on the board. This proved to be a good arrangement and profits were split on a patronage basis.

CSP FOODS LTD.

During the Second World War an entrepreneur, Gordon Ross of Moose Jaw, approached the Saskatchewan Wheat Pool and asked for help to find farmers to grow a crop called "rapeseed." Ross had a small oil seed crushing plant and had experimented with rapeseed. The oil, while not edible, had properties that were desirable, as oil for marine engines, particularly submarines.

Ross had been contacted by the government of Canada and was asked to supply as much oil as possible as part of the war effort. Needless to say, the Saskatchewan Wheat Pool took up the task and soon had enough supply to run his crushing plant to capacity. Following the war, the urgency to supply oil was gone, but the Saskatchewan Wheat Pool saw the commercial potential of this new crop. The oil also was desirable as a lubricant in the cold rolling of steel.

The outcome was that the Saskatchewan Wheat Pool purchased Ross's crushing plant and moved it to Saskatoon in 1947 as the start of Saskatchewan Pool's industrial activities. The following year, a flour mill was built beside the vegetable oil crushing plant. Farmers were eager for new crops and readily took to growing rapeseed. It was compatible with wheat in crop rotation cycles, required more moisture than wheat in the growing period, but was more durable than wheat in the often wet harvest conditions.

Many agrologists felt that greater use could be made of this crop. The Saskatchewan Wheat Pool, working with the universities in Saskatchewan and Manitoba, began breeding programs to achieve an oil that was edible for humans and safe for animal feed. Over a period of years, scientists dedicated to this venture were successful. Roy McKenzie, the director of the Pool's Farm Service Division, encouraged his staff to handle this project. During our summer, they would grow a crop in Saskatchewan that was cross-pollinated to obtain a certain feature. Then, rather than wait until the following year to further upgrade the seed, he arranged to grow a winter crop in California and in Chile. The Saskatchewan Wheat Pool and others invested a great deal of time and money in this project, but it paid off handsomely for prairie farmers in the successful development of a new crop known as "canola"; the name was changed to avoid the unfortunate connotations of the name "rapeseed."

The Gordon Ross plant that had been moved from Moose Jaw was replaced with a larger, state-of-the-art plant and the business of selling oil and meal was undertaken with great vigour. There was resistance for some time by the makers of margarine, but they soon came to recognize the value of canola oil, and meal moved freely into feeding operations. It was obvious if we were to meet the growing demand that expansion was necessary. Sites for a new plant were being explored when the Saskatchewan Wheat Pool had the opportunity to purchase a crushing plant at Nipawin. The purchase was made and proved to be very successful. It was located in the centre of the best canola-growing area and it complemented our Saskatoon operation. The Saskatchewan Wheat Pool oil and meal markets expanded dramatically and there was potential for even greater quantities of each to be sold.

The Alberta and Manitoba Pools were not in the canola business but were being pressured to set up plants. United Grain Growers, a competitor, opened a plant at Lloydminster and this put further pressure on Alberta Pool. Manitoba Pool Elevators bought the Co-op Vegetable Oils facility at Altona, Manitoba. It was a good match because CVO needed to expand but lacked the capital to do so. They were pleased to see their plant sold to another co-operative group.

There was still the potential for great expansion, so the three Pools decided to form a joint company. The Saskatchewan Wheat Pool would contribute its plants in Saskatoon and Nipawin, Manitoba Pool Elevators their plant in Altona, and the new company—CSP Foods Ltd.—would build new plants in northern Alberta and in Harrowby, Manitoba. The night before the agreement was to be signed, Alberta Pool withdrew, opting instead to form a partnership with a large Japanese grain company and to build a new crushing facility in Alberta. Nevertheless, the Saskatchewan and Manitoba Pools signed the agreement and, in due time, the Harrowby plant was constructed. I was privileged to be on the board of CSP Foods Ltd. for several years and was delighted by the progress the company made. This was another fine example of seizing the opportunity for joint action.

PRINCE RUPERT GRAIN LTD.

The turmoil in the whole grain area that was marked by the Crow Rate debate also sparked an examination of the whole grain-handling system. As I indicate elsewhere, the Saskatchewan Wheat Pool made a conscious decision to reduce the number of country elevators and thus provide system efficiencies.

Export volumes of grains were increasing due to companies like XCAN Grain and the stalwart efforts of the Canadian Wheat Board. The major new markets were in Asia—Japan, China, and other countries that were easier to serve out of Canada's Pacific ports. Since the Common Agriculture Policy of Europe was determined, at all costs, to reduce Europe's reliance on imports of grain and did so by subsidizing production, Canada was thus confronted with declining export volumes to Europe. Canada's export infrastructure was heavily concentrated in Thunder Bay and the St. Lawrence Seaway, and it was evident that there was no need for expansion in facilities to assist the eastern movement out of Canada.

The Saskatchewan Wheat Pool built a new grain export terminal in Vancouver that opened in 1968, but even with that new plant and access to the purchased Federal Grain terminal, we would be hard-pressed to meet our shipping opportunities. Prince Rupert, some 500 miles north of Vancouver, was home to a rather small and antiquated grain terminal that was operated by the federal government. Because Prince Rupert was a full day's sailing closer to Japan, farmers would have a net benefit because of lower ocean freight charges. Many companies were considering the plant with the view to acquiring it, but were unsure of how much money would be required to update it. Railway grain cars were also in heavy demand, and it was uncertain if an adequate supply would be available for an expanded Prince Rupert port facility. As well Prince Rupert was only serviced by Canadian National Railways (CNR), thus limiting, the collection area for grain to only CNR lines on the prairies. These factors combined to scare away prospective purchasers.

One day, I received a phone call from Roger Murray, president of Cargill Canada. He knew I was going to Winnipeg and asked if we could meet for lunch. At our meeting he suggested that the Saskatchewan

Wheat Pool and Cargill jointly purchase the Prince Rupert terminal. Cargill, which was heavily concentrated on the St. Lawrence Seaway, was not well-positioned for the shift to grain exports out of Pacific ports. I could think of no good reason why we would go into business with Cargill, before first exploring the opportunity with the Alberta and Manitoba Pools. So, I told Roger we had no interest in a two-company joint venture, but would participate in an industry approach.

In the meantime, the Alberta government was anxious to get involved, or at least influence matters, in the grain industry. It successfully cajoled the major grain industry players to come together and to consider Prince Rupert as a site for a new export grain terminal, and so the Prince Rupert Grain Terminal consortium was formed. The members were the three Pools, United Grain Growers, Pioneer Grain—which had a fairly new terminal in Vancouver—and Cargill Grain. The Alberta government was represented by Deputy Premier Hugh Horner. Allan Macpherson of Alberta Pool was selected as chair of the consortium. The initial meeting was held in 1982 and the official opening of the terminal occurred in 1985. The three-year interval was filled with numerous meetings which increased in importance once the decision was made to proceed. Milt Fair, CEO of the Saskatchewan Wheat Pool, and I were on the board and Michael Thompson, chief engineer of the Saskatchewan Wheat Pool, became the construction guru.

It was a difficult challenge and if not for the pressure of Hugh Horner, may not have been accomplished. Horner committed Alberta to considerable financial support for the construction in the form of loan capital. The individual companies had to determine if they could afford the project and how it would affect earnings of their present facilities. Cargill was eager because it was inadequately served on the Pacific, but did not originate sufficient volume to go it alone. The federal government was committed to the project but not by way of financing. There was no doubt we could build the $300 million plant, but it had to be a viable operation.

The CNR gave assurances of its ability to move the required volumes of grain to the port, but even at full capacity the return on investment did not meet the project standards most of the companies required. We were doing careful analysis on how we could be assured of viability and concluded that we required $2.50 per tonne additional revenue to do so.

Prince Rupert Grain Board planning session. Left to right: Mac Runciman,
United Grain Growers; Gordon Harold, Alberta Wheat Pool; Bruce McMillan,
Pioneer Grain; and, Ted Turner, Saskatchewan Wheat Pool.

I suggested that we seek permission from the Canadian Grain Commission, which controlled handling charges, to allow us to increase our charges by the required amount. Eugene Whelan, the minister in charge of the Grain Commission, indicated verbally that a $2.50 per tonne increase would be allowed, so we made the decision to proceed. It should be noted that because of closer proximity to Asia, the higher handling charge would be offset by lower ocean freight costs.

After we committed, the minister withdrew his permission and we were stuck with an uneconomical operation. It was quite a few years before the terminal became viable. While the owners were stuck with a less than remunerative operation, the grain industry benefited from increased capacity. Personally, I found it enjoyable to be working with our competitors on this undertaking.

RELATIONSHIPS

Over the years, the Saskatchewan Wheat Pool developed many important relationships. Some were formal, while others were at the convenience of either party. Three relationships were especially significant: our involvement with trade unions, with the government of Saskatchewan, and with the government of Canada.

Trade Unions

Within the broad scope of the Saskatchewan Wheat Pool, we dealt with several unions at Thunder Bay, Winnipeg, Vancouver, Regina, and Saskatoon. For the most part, communications were between the unions and our human resources personnel. However, at contract time, the corporate offices of the Saskatchewan Wheat Pool became involved.

Our largest union was the Saskatchewan Wheat Pool Employees' Association (SWPEA)—affectionately referred to as "Sweet Pea"— which later became the Grain Services Union (GSU). For the most part, we had excellent relations with the unions. Our senior people were not opposed to unionism and dealt openly and fairly with the collective bargaining units representing the employees.

As first vice-president, I chaired the negotiating sessions with the GSU and on several occasions the negotiations were protracted. The head of the GSU was a crusty Scot by the name of George Mills. He had been an elevator agent of long standing and knew the grain business thoroughly; he was also a dedicated unionist. He was a tough bargainer, and pushed the Saskatchewan Wheat Pool very hard on contract items. George, like virtually all of the country elevator agents, had a real passion for the goals of the Saskatchewan Wheat Pool, and while seeking the best for his union members, he was always cognizant of the need for the Pool to be a viable organization. He reasoned that his members were best served by a strong Pool that could afford to maintain jobs and pay remunerative wages.

Following Mills's retirement, the GSU hired Bill Gilbey as general manager. The contrast between the two men could not have been greater. Gilbey never had a good word to say about the Saskatchewan Wheat Pool and used the negotiating sessions as an opportunity to bash the employer. The consequence was that the Saskatchewan Wheat Pool executives fought back by granting only those requests which were clearly in the best interest of the company. Going into negotiations, the executive had clear goals on what we wished to achieve. George Mills, with his consideration, came closer to achieving the maximum than did Gilbey, with his combative approach. I always felt there must be a better way than a confrontational meeting where participants were forced to take extreme positions. Management could not place on the table what it was prepared to do because that then became the minimum, and negotiations would go on from there.

The GSU was in a tough position to force a strike because the majority of its members were country elevator agents, so in many cases, its members would be striking against their best friends in the rural communities. Nevertheless, there was a strike at Thunder Bay and strikes at Vancouver, where we dealt with the Longshoremen's Union and we bargained as an industry. Their demands were often quite unreasonable, and they timed strike action to coincide with periods of heavy grain movement. The result was a line-up of ships waiting to be loaded with grain and demurrage payments being accumulated, which caused a great outcry in the farming communities. The pressure on grain companies, including the Saskatchewan Wheat Pool, was tremendous, as we were pushed to make a non-economic settlement. During one strike in the mid-1960s in Vancouver, Prime Minister Pearson intervened and told the companies to agree to the 30 percent wage increase being sought by the union. This touched off a wave of unwise settlements across the country and sparked a round of inflation. Henry Kancs, the union manager, had no understanding of the needs of farmers and cared little that the union was placing a financial burden on them.

Generally speaking, however, our labour experience was good, considering the number of opportunities for labour strife. This was a credit to our division managers and to the alertness and skill of our Human Resources Division.

The Government of Saskatchewan
The Government of Saskatchewan always had a healthy respect for the Saskatchewan Wheat Pool. This was because we were very influential in Saskatchewan and had connections in every area of the province, more so than members of the Legislative Assembly. For the most part, we worked quite well together, regardless of the party in power. As president, I would meet with the premier on one or two occasions each year, at which time either participant might initiate the meeting. These private chats allowed for a good understanding of the issues facing each other.

On one occasion, just prior to the 1982 provincial election, NDP premier Allan Blakeney brought his cabinet personnel to meet with our board of directors. Led by Gordon McMurchy, minister of agriculture, the government members warned that they would be making the Crow Rate part of their election platform and would thus be criticizing the Saskatchewan Wheat Pool. I warned that they should be careful in that they might not get the reaction they were looking for. In the ensuing election, the NDP was defeated and Grant Devine headed a Conservative government.

The premier of our province was always asked to speak at our annual meeting of delegates, as was the minister of Agriculture. When pressing Ottawa for action on agricultural issues, we always first sought to enlist the support of the provincial government. Overall, we had a very positive relationship with provincial government officials during my tenure as president.

The Government of Canada

The Saskatchewan Wheat Pool had an uneven relationship with the federal government over the years. Throughout my 25-year tenure on the executive of the Saskatchewan Wheat Pool, there was never a time when there was not active dialogue or some type of venture that involved the two entities, but that dialogue ranged from being harmonious to hostile. At times, government officials were quite flexible and open to receive suggestions, most likely when they were developing a policy position. At other times, they were completely unapproachable; an indication that their minds were already made up and would not be changed. Assistants to ministers were generally, it seemed, more interested in gaining "brownie points" from the minister than in making changes in policy, even though they could see that farmers would benefit.

Ottawa was the focus of most of the resolutions that were adopted by the annual meeting of Saskatchewan Wheat Pool delegates, so it was only natural that we would be in frequent contact with some minister or department of government. I met individually with Prime Ministers Trudeau and Mulroney, and I met with numerous cabinet ministers, establishing excellent rapport with such people as Eugene Whelan, John Wise, Don Mazankowski, Jean-Luc Pepin, Ralph Goodale, and John Crosby.

I served on numerous advisory committees encompassing economic and trade-related matters. The Saskatchewan Wheat Pool participated on Canadian International Development Association (CIDA) projects in Africa and also in the Philippines. Negotiating international grain agreements found us very much on the same page and we strove together to achieve a good deal for Canada.

CHAPTER 7

The Canadian Wheat Board

The Canadian Wheat Board was established as an instrument of public policy during the First World War and made permanent by legislation in 1935. Like any public policy creation, it was not unanimously embraced by those most directly affected, nor even indeed by those who stood to benefit the most from such an entity. Because of opposition to the CWB, at the annual meeting of Saskatchewan Wheat Pool delegates felt it necessary to pass a motion, each year, in support of the CWB; the motion was most often carried unanimously.

The government quite wisely decided to have an arm's length relationship with the CWB rather than making it a department of government. A minister of the crown was named as "Minister Responsible for the Canadian Wheat Board." This designation was in addition to a regular portfolio identification and duties. It was usually bestowed upon the minister of agriculture, the minister of industry, trade and commerce or the minister of transportation. This link from the government to the CWB was more than communication but less than direct supervision, and allowed the government to be supportive without being involved in the intricacies of CWB operations.

There were three main areas where the government was wonderfully supportive: the most frequent and visible was a financial obligation. Prior to the start of the new crop year on August 1 each year, the

CWB set an initial price for grain. This was the price farmers would receive upon delivery of their grain to a country elevator facility. The initial price was determined by an assessment of global market conditions. The government by cabinet decision would underwrite the initial price. On only a few occasions was the government called upon to make up a deficit resulting from the initial price being higher than the market generated selling price; also the government either directly or through its worldwide network of offices would help in the marketing function. The most notable example of this was in 1961–62 when the minister of agriculture, Alvin Hamilton, and CWB officials made a huge wheat sale to China. Alone, the CWB had not been able to approach China because Canada did not officially recognize China at that time. This coup was of lasting benefit to Canadian grain producers; the government was also very helpful by providing protocol to state trading countries such as China, the USSR, and Japan. The minster responsible would host visiting grain delegations and/or take delegations on overseas promotion trips. Hazen Argue and Charlie Mayer were very proficient in this role.

Canadian grain executives hosting a delegation from the USSR in 1985.
Among the Canadians were: Ted Turner, President of the Saskatchewan Wheat Pool
(2nd from left), Esmond Jarvis, Chief Commissioner, Canadian Wheat Board
(5th from left), and Agriculture Minister Hon. Charlie Mayer (8th from left).

The CWB established a pool for each grade of every grain handled. The marketing experience for each grade was recorded and when the pool was closed—usually on July 31 the following year—the difference between the realized price and the initial price was paid to farmers on a pro-rata basis commensurate with the number of bushels they delivered of each grade of grain. When the final payment cheques would arrive there was great excitement in the community and a noticeable stimulation of economic activity. The final payments always attracted significant media attention, and when some journalists would describe them as a government handout, it was then incumbent upon me and other farm spokesmen to set the record straight—that the money resulted from a well run marketing entity. All the operating costs of the CWB were paid by farmers as a reduction of their final return of the pooling accounts. The CWB allowed Canada to counter the policies of other countries, especially those of the United States, where the Department of Agriculture freely utilized public policy to influence the production and marketing of farm commodities.

The CWB was central to the prairie grain-producing area. All grain companies signed a handling agreement with the CWB and were paid for receiving and storing CWB grains. This was a major revenue source for each of the companies involved. To safeguard the integrity of the system, the handling and storage charges were set by the Canadian Grain Commission, which negotiated the charges with the handling companies.

By 1975, the grain glut of the early years of the decade was over, and there now loomed the potential of a major wheat shortage, with a correspondingly dramatic increase in prices. Increased demand emphasized the need for a free flow of grain to port positions. In 1968, the Saskatchewan Wheat Pool had opened a new port terminal in Vancouver and it was being used to capacity. Grain cars and railway motive power were proving to be weak links, which prompted the Wheat Board, and the federal and provincial governments to purchase hopper cars, and the railways to lease additional motive power.

The federal government was threatening to take over the industry and to force changes in grain-handling systems, while the railways were abandoning rail lines to improve their efficiency. Numerous studies were being conducted; all, it seemed, recommended reducing the mileage of railway collection systems and blamed the Canadian Wheat Board for every problem.

Otto Lang, either deliberately or inadvertently, had brought criticism of the Canadian Wheat Board to the forefront. This was vigorously picked up by the private grain trade and the newly formed Palliser Wheat Growers. There was a good deal of dissatisfaction among farmers, as disruptions in the grain delivery system resulted in missed export opportunities.

PERSONAL INVOLVEMENT WITH THE CANADIAN WHEAT BOARD

From my earliest days as a youth, I was aware of the important role played by the Canadian Wheat Board in the agricultural industry. Yet, despite its importance, I knew that there were those who opposed the Board. One source of opposition sprang from a deeply held belief that the Canadian Wheat Board interfered with the free-enterprise system. Certainly that is correct, but for good reason—the free enterprise companies were selfish and failed to meet the needs of farmers. Others opposed the Board because in the bleak days of the First World War, it served to keep the price of wheat lower than would have been the case otherwise. The market was Great Britain, and it deserved to receive whatever help could be sent its way. Those who protested should have directed their wrath at the Canadian government, arguing that the difference—perceived or otherwise—should have been paid to wheat producers, and that such action for the public good was the responsibility of the country, not only of farmers. Among Wheat Pool members, there continues to be overwhelming support for the Canadian Wheat Board, and at every annual meeting of delegates, a resolution is passed stating unequivocal support for the Board.

In my opinion, the record of the Canadian Wheat Board has been stellar. Like every other successful business, not every transaction was outstanding, but in dealing with state-trading countries like China and Russia, the Canadian Wheat Board was in a league of its own. My first experience in dealing with the CWB was when Bill MacNamara was the chief commissioner. Bill had been a Saskatchewan Wheat Pool employee and so there was no shift in philosophy. However, many of the subsequent commissioners and chief commissioners came from the private sector and, without exception, became passionate in their commitment to the functioning and philosophy of the Board.

To cite a few "converts" to the Canadian Wheat Board: Chief Commissioner Gerry Vogel, who had been the sales manager of Bunge; and, Esmond Jarvis, a behind-the-scenes civil servant who, at one time, opposed the Board, also became a passionate chief commissioner. At the time of his appointment, I wrote a letter strongly opposing it. I sent him a copy. Over the years, we became close friends and we often laughed about the letter. He held no resentment because he understood where I was coming from. Forrest Hetland, while president of the Saskatchewan Canola Growers' Association, fiercely opposed the Board. Forrest was appointed to the Canadian Grain Commission as a commissioner and later as a Canadian Wheat Board commissioner. He became totally committed to the Canadian Wheat Board. Lorne Hehn, during his time as a director and as president of United Grain Growers', was not supportive of the Board. He and I would have sparring matches in agricultural forums about the Canadian Wheat Board. As chief commissioner of the Canadian Wheat Board, he became one of its strongest advocates, and did a good job of refuting most of the criticisms that were hurled at the Board. What better testimony about the worth of the Canadian Wheat Board than the actions and dedication of one-time opponents?

THE BATTLE TO CHANGE THE MANDATE OF THE CANADIAN WHEAT BOARD

In the days when Otto Lang was the czar of Canadian grain matters, he became determined to weaken the Board or to destroy it completely. This, in turn, became the major cause of prairie farmers' disenchantment with him. His action inspired others to express opposition and so a public debate started that continued for years. Much of this was fostered by companies closely aligned with the open-market philosophy. Others took a narrow view of the marketplace and believed that they could, on their own, get a higher price and adequate market for all of their own production. Generally, they quoted some American point as paying more for a specific grain than the CWB on a specific day. They failed to recognize that if any significant quantity was moved to that market, it would quickly reduce the price and/or the United States would embargo sales to protect its own farmers.

The battle raged as many Conservative politicians joined in the criticism, even promising to eliminate the Canadian Wheat Board entirely. They ignored the fact that in most elections of the Canadian Wheat

Board Advisory Committee, pro-board candidates won over anti-board candidates. The traditional opponents of the CWB have been private grain companies such as the now defunct United Grain Growers, the Western Wheat Growers (formerly Palliser Wheat Growers), the Barley Growers' Association, livestock groups, certain departments of the University of Manitoba, the United States federal government and certain state governments, the Chamber of Maritime Commerce, the Thunder Bay Harbour Commission, the Conservative Party of Canada, and, more recently, Viterra—even though the three prairie Pools (now encompassed by Viterra) had traditionally supported the CWB. This is a clear indication of how far Viterra moved away from the grassroots policies of the Pools, due to lack of contact with other than a selected few farmers, picked because they would support the open market philosophy of Viterra management.

The strongest advocates of the Canadian Wheat Board are the National Farmers' Union, the Liberal Party, the New Democratic Party, and select provincial governments, including Quebec. Quebec views any attack on the CWB as an attack on marketing boards in general, which have been key to a successful agricultural industry in the province.

I served on the Canadian Wheat Board Advisory Committee for 13 years as an appointed member and an elected member. Never at any time was I disappointed in the approach or the efforts by either staff or commissioners. They served Canadian farmers and Canada very well indeed, and were highly respected around the world, especially by our largest customers for wheat and barley. Farmers still are supportive of the CWB, as was seen in the 2010 Advisory Committee election where four out of five successful candidates were pro-Board supporters.

Since it came to power in 2006, the Stephen Harper Conservative government has stated its intention to remove the single-desk feature of the CWB. While Harper headed minority governments he was unable to restructure the CWB, because it was impossible to pass legislation that would significantly change the Board's mandate. However, the Conservatives won a majority in the 2011 federal election, and announced that they would reduce the powers of the Board, to "give choice" to farmers in marketing their grains.

Conservative agriculture ministers have repeatedly stated that the CWB will continue to exist but have not stated how or in what form.

The CWB does not own grain-handling facilities, and functions by paying companies for services rendered. This works extremely well while the CWB is in a monopoly position, but if the handling companies are granted the right to manage the export of wheat and barley, the Board will then effectively be reliant on its competitors for services. While this is possible, it requires a contract to pay for the required services and there are many uncertainties about such arrangements. In the initial days of the Saskatchewan Wheat Pool the case was identical—the Pool did not have facilities and engaged the elevator companies of the day to handle "Pool" grain. Often a producer affiliated with the Pool would arrive at an elevator with a wagonload of grain, only to be told that the elevator lack the space to take "Pool" wheat. It was not until the Pool purchased the Cooperative Elevator Company facilities that progress was made and it began to expand rapidly. Similarly, export terminals were not reliable until the Pools owned or controlled them outright.

After the 2011 general election, the federal government claimed that it had a mandate to alter or terminate the CWB. Many farmers disagree, and a poll of grain producers initiated by CWB—and conducted by Meyers Norris Penny—indicated over 60% favoured retaining wheat and over 50% retaining barley within the jurisdiction of the CWB. The federal government and several other groups have denounced the validity of the poll.

It is revealing that the federal government refused to conduct a plebiscite of grain producers on the future structure of the CWB, which suggests that it fears farmers would vote to retain the Board. Instead the government chose to allow those not directly involved in growing or selling grain to determine the fate of the CWB.

Legislation decreed that on August 1, 2012, the monopoly powers of the CWB would cease, and producers could thereafter sell their grain wherever they chose. This ended the single-desk selling function that had served Canadian farmers well for many years. A year later, the jury is still out on the impact of the change. The first year saw adequate markets and prices higher than the previous year. Critics of the Board are claiming this phenomenon is due to the removal of the CWB. It is more likely that a poor crop in the United States was responsible for reduced world grain stocks and hence higher prices.

Prior to the actual decision to remove the single-desk selling function of the CWB I raised three questions, with related comments, and one possible solution that could have potentially satisfied each side of the Canadian Wheat Board debate. However, the government was committed to ending the single-desk selling feature of the Board and could not be swayed from that goal. My questions were:

1. Why is the Canadian government keen on destroying what is an entirely Canadian entity?

2. Why is the Canadian government anxious to negate this effective central selling agency, and thus accomplish what the American government has been trying to do for many years? The United States on nine separate occasions has challenged the CWB under international trade laws and has lost every time. Is it realistic to believe that American jurisdictions have the best interest of Canadian farmers as a priority?

3. Was the Canadian government prepared to subsidize farmers at the same relative level that the United States subsidizes its farmers? The American system is what will remain if the CWB is removed. The United States has relied on subsidies to attain viable farm operations under its system. Within the CWB system, the only exposure for Canadian taxpayers is the initial price determined by CWB and guaranteed by the federal government. On only a handful of occasions since 1935 has the initial price been higher than what was received from the marketplace. Thus, the cost to government has been minimal. There have been times when government payments to grain producers were required to offset the impact of European and American export subsidies, but the CWB system itself was not a burden on the Canadian treasury.

A POSSIBLE SOLUTION

I believed that a compromise could have been found that would have met the objectives of each side of this intense debate by the following:

1. Leave the CWB intact: with its single-desk selling functions, initial price setting responsibility (guaranteed by the government of Canada), ability to set quotas and with access to country elevators and export terminals, that are declared "works for the general advantage of Canada."

2. Farmers would not be obliged to use the CWB system and could opt out for a period of not less than five years. Similarly, those wishing to use the CWB must declare so and commit to at least five years of participation.

There are advantages to both CWB supporters and those favouring the open market system. Identifying the acreage committed to each system would allow the CWB on one hand and also the private marketers to assess the amount of grain they would have to market and they could forward-sell accordingly. The five-year rolling commitment would allow them to enter into long-term contracts with purchasers. The net result would be a smaller but still viable CWB and an open market availability. I am confident that, while complex, the details could have been worked out.

CHAPTER 8
The Globalization Phenomenon

The settlement of the prairie region of Canada resulted in rapidly developing farm enterprises. The rich land and the dedication of farmers soon produced surplus grain beyond Canada's consumption needs. As grain production, particularly that of wheat, increased, prairie farmers relied more and more on exports to maintain viable operations.

Canada had a distinct disadvantage compared to most other surplus-producing countries, since most of the grain produced in Canada is far from water transportation. On Hudson Bay, the port of Churchill is the closest to the prairies but has limited export capacity and is only operational for three months each year. Thunder Bay is about 1,000 miles to the mid-point of the prairies, but still 1,500 miles to tide water, and only operational nine months of the year. The Pacific coast is also about 1,000 miles to the mid-point of the prairies and is located on tide water, but is separated from the prairies by the formidable Rocky Mountains.

In Australia and Argentina, most of the grain is grown within 300 miles of tide water. Europe's river system reaches well into the continent and the distance to tide water rarely exceeds 500 miles. In the United States, the Mississippi River services the grain-growing area, and all of the United States waterways are heavily subsidized.

It became obvious that Canada's logistical disadvantage needed to be offset with a superior marketing structure. During the First World War, in order to keep the price of wheat from going "through the roof," the Canadian Wheat Board was established. The Board paid an initial price to producers and returned profits by way of a final payment.

While farmers were not happy about the dampening of grain prices, they accepted it as part of the war effort; and, they did like the price-averaging feature of the system. Previously, the price was lowest at harvest time when farmers had to sell to pay debts; then the price rose later in the year as supply weakened. Those who could afford to wait to sell were usually rewarded by a higher price.

The Canadian Wheat Board was intended as a temporary measure, but clamouring by farmers kept it in operation and in 1935 the federal Conservative government introduced legislation making it permanent. The performance of the Canadian Wheat Board in subsequent years was outstanding, much to the dismay of other countries that did not have such an agency. The Unites States brought complaints against the CWB under the General Agreement on Tariffs and Trade on nine occasions, only to see the CWB practices upheld every time.

INTERNATIONAL GRAIN WARS

Following the Second World War, most exporting countries found that they had surpluses in agriculture products and looked beyond their borders to dispose of these products. Meanwhile, importing countries sought to increase their production and thus reduce their reliance on imports.

To recover from the economic devastation of the war, countries in Europe collectively formed the European Economic Community (EEC). One of their main objectives was to ensure an adequate supply of food. To enhance production they established a Common Agriculture Policy (CAP) which provided for heavily subsidized production, encouraged surpluses—the cost of which were carried as an EEC expense—and put up prohibitive barriers against food imports. To this day CAP policies distort trade.

The United States made an aggressive effort to blast its way into world wheat markets by introducing Public Law 480, which allowed American companies to sell below world price levels. This bill protected American producers through a system of floor and target prices, so American farmers were always assured of getting a remunerative price, while devastating farmers in other countries. As well, the bill allowed food aid provisions without damaging the farmers, and wheat was provided to recipient countries in exchange for American access to their strategic military sites. These provisions were applied to regular customers of Canada, Australia, and Argentina. The American transportation system featured subsidized waterways so the United States was selling into world markets below the cost of production in other countries while holding its farmers harmless.

Because of such action, Canada struggled to hold traditional markets and found it difficult to develop new ones. However, early in the 1960s, Canada was able to sell large quantities of wheat to China and the USSR thanks to the Canadian Wheat Board and the work of federal Agriculture Minister Alvin Hamilton. These sales, at remunerative prices, saved the grain industry in western Canada. The United States was livid that Canada would sell to communist countries, but American farmers quickly lobbied their government for permission to sell into this lucrative market and soon got permission to do so. It was interesting how quickly the American government set aside its philosophy of not dealing with communists when its own farmers were being hit in the pocketbook. However, because of American foreign policy towards China and the USSR, and because those countries' needs were being well-served by Canada, they showed little interest in the United States as a supplier.

In 1972, the USSR experienced a mammoth crop failure and required more wheat than Canada was able to supply. A delegation from the USSR went to the United States to purchase the wheat it required to meet the needs of its country. In the United States, the Soviet delegation met with each of six large grain companies—Cargill, Continental, Cook, Dreyfus, Bunge, and Garnac. It negotiated quantity, price, and terms of delivery—normal contract provisions. However, each company was under the impression that it was the only one dealing with the Soviets. The six companies were appalled when they went to the production area to secure the grain they had each committed, only to find themselves in competition with each other.

Farmers soon realized that wheat was a hot commodity and the price soared. The end result was that the exporting companies had to pay more for the wheat than their committed price to Russia. There are huge financial penalties, as well as reputation concerns, for reneging on a contract, so the six companies had to swallow the loss. The Americans cried "foul" and Russia was vilified, but nothing illegal had been done and the losses were real. Interestingly, the United States' government bailed out the grain companies. Various reports placed the sum at a cost of between $130 million and $400 million.

The following excerpts are from the book, *The Great Grain Robbery* by James Trager, a revised and updated version of *Amber Waves of Grain*:

> Actually, the USA exporters had been the recipient of export subsidies for several years. Not only did they receive the difference of what they paid producers and the price they received from importers, they also received a tax break for doing so.

> It is also a matter of commercial self-interest. The main emphasis of the 1954 law (Public Law 480) was never on helping the economic development of recipient countries in the first place: it was on disposing of US surpluses whose producers had no market and were threatened with falling prices. Only commodities in surplus could be shipped under Public Law 480. So, while butter, cheese, dry skim milk, and vegetable oils were donated or sold for native currency ("funny money," it was called), in the earlier years of the program. As US stocks of those foods declined in the 1960s, almost the only remaining Food for Peace was wheat.[1]

> One thing was certain. However efficient US agriculture may be, and whatever that efficiency may owe to a competitive free enterprise economy, it was the free market system that enabled the Russian purchasing agents to buy so much US grain without having Americans find out about it. Had the deals been made with Argentina, Australia or Canada, the Russians would have had to negotiate with government monopolies, they would have been unable to fragment purchase among six private firms, with none of the six knowing what the others had sold, and with the US Department of Agriculture too long in the dark about the entire matter.[2]

1 James Trager, *The Great Grain Robbery* (New York: Ballantine Books, 1975), 133.
2 Ibid., 159.

Since the exporters had been granted $300 million in subsidies on the wheat they had sold to foreign nations, including the USSR, the Treasury on October 4, 1972, issued tentative regulations denying them the special tax. Continental, Cargill, and the other export companies protested. They said the law, in its reference to exports already subsidized, was about concessional sales such as those made through the Agency for International Development: it was not referring to sales made on ordinary commercial terms. Cargill argued that foreign trade was subsidized in a variety of ways, that to merchandize US wheat successfully, an export company needed both export subsidies and tax benefits. Supporting this position was the Chairmen of the Senate Finance and Agriculture Committees.[3]

However, the feeling lingered that Russia had somehow cheated the United States and there was latent anger in grain circles.

The following year, I was invited to speak to a grain symposium in New York City. It was a high-level meeting with senior grain executives from all over the United States. At the onset of the meeting, on a point of privilege, a request was made to have someone from Russia come to the meeting to account for the deception that Russia had used. It was announced to the meeting a short time later that a senior Soviet official would attend at the start of the meeting on the second and last day of the conference.

I had made my remarks on the first day, so I eagerly awaited the appearance of the Russian diplomat. Upon taking the stage, he acknowledged that he was aware of the upset in the United States about the grain deal. He denied, however, that the USSR delegation had implied that it was only talking to one company (each company had said, to cover its actions, that it had exclusivity for the deal). There was silence in the room when the Russian said, "I thought you would be proud of us because we used the American way." I had to muffle my laughter.

INTERNATIONAL WHEAT/GRAIN AGREEMENTS

One of the major objectives of the Saskatchewan Wheat Pool was to achieve remunerative and stable prices for Canadian grain producers. To that end it had for many years promoted the idea of an international

---

3 Ibid., 160–61.

wheat agreement. The agreement would, if successful, provide a fair return to producers by way of a remunerative floor price. It would also provide stability in world markets because importing countries would be assured of supply and the protection of a ceiling price.

I accepted an invitation to advise at a negotiating session of an International Wheat Agreement at Geneva in 1971 and 1972. The conference was unsuccessful in large part because the exporting countries could not come to a common understanding; in particular, the United States did not want an agreement that might limit its ability to subsidize the production and export of wheat. The conference came close in November 1971 to achieving a price range that was fair to producers, a stock-holding policy to provide for emergency situations in which the burden would be shared by importers and exporters, and a commitment to food aid, also shared by importing and exporting countries. However, when the IWA reconvened in February-March 1972, it was soon evident that such was not achievable. The importing countries had experienced and were enjoying the prices that resulted from heavy world wheat stocks, and the United States, through subsidy encouragement, was expanding its export volumes and production base. The United States proposed that a single price be set at about our floor price suggestion and that would then be the maximum. Canada refused, and everyone went home. I met former Prime Minister John Diefenbaker on an aeroplane and he complained to me about not getting an agreement. My response was, "I don't need an international agreement to give me a lousy price; the market will do that at any time."

Canadian Wheat Board commissioners were in attendance and proved to be invaluable with their world market knowledge and how they conveyed it to the Canadian delegation. I was startled to realize that the biggest obstacle to overcome was within the exporter group—I had gone to Geneva expecting it to be the exporters against the importers. I was also surprised by the scope of the attendees. I had expected a few exporters—perhaps five or six—and many importers, perhaps represented by twenty of the largest importing countries. Also attending were many developing countries, even as small as Malta; the developing countries were very vocal and influential.

I was dismayed by the approach taken by the United States. I had anticipated that it, too, would want a remunerative floor price for its producers. Not so. The Americans advocated a ceiling price about where we felt the floor price should be. The United States was the only major exporter to take that position and, not surprisingly, it was embraced by importing countries. I was convinced that the policy of the United States Department of Agriculture was to keep the price so low that it would force other countries out of the market. This was the topic of discussion for a panel made up of Les Price from Australia, an American representative, and myself at a meeting in Denver. The American position was clearly outlined, and I reacted to it as follows:

A Canadian wheat growers' representative (E. K. Turner) told US growers, "Now, you know the arguments they (USDA) put up. If you get the price down, you'll force others out of the market, and leave an enlarged market for your own products.

"I ask the question: Where is this going to happen? It certainly is not going to happen in one of our three countries (Canada, Australia and the United States). It's not going to happen in the European Common Market, because, for the most part, the producer doesn't even know what the world price is. He's not exposed to it. It's not going to happen in the developing countries, because they are striving for self-sufficiency anyway, and they simply don't have the means to go and purchase the grain that they need to feed their people. So I ask: Where is this going to happen, simply by lowering the price?

"What advantage is it to the producer to sell 10,000 bushels for $10,000 instead of selling 8,000 bushels for $10,000? But it's sure a heck of a good thing for the handling companies and exporters and everybody that's in between the farmer and the market, because they work on a volume margin. What is the advantage to the balance of payments in selling more wheat for the same number of dollars?"[4]

4 Marsha Hamilton, *The Great American Grain Robbery: An Agribusiness Accountability Project Study.* The specific remarks were lifted by Hamilton from the speech I made to the National Association of Wheat Growers (NAWG) in Denver, January 1972, 81–82. The Regina Public Library brought this study in for me and I was only allowed to keep it for a few days. Imagine my surprise when perusing Hamilton's report for the United States Senate to discover that I was being quoted.

Pressure for putting floors into the agreement built up substantially, though, with no apparent effect on the USDA. Canadian, Australian, and Common Market wheat grower representatives urged American growers to mount pressure on the United States government for a floor price in the agreement. Being aware that the American Senate had passed a resolution calling for a floor price, I made the following comment:

> The US has the Senate resolution for impetus and you're the dominant nation in the world wheat trade and marketing. These are good reasons why I think the United States ought to take the lead in calling for a new agreement.[5]

Pressure from Canada and Australia, and the Senate resolution had no effect on the USDA, which used the large grain exporters as advisors on such matters. Since the American producers were shielded from the world price, they put up no outcry.

Under the GATT in 1977 and 1978, we made one more attempt to achieve an International Grains Agreement (IGA). It had been concluded that to get an agreement, the base coverage needed to be broadened. The resulting conference was almost a carbon copy of the 1971–72 efforts. In summary, I was impressed and pleased with the position and efforts of our government officials; they tried hard and listened to producer representatives.

THE NORTH AMERICAN FREE TRADE AGREEMENT

In 1984, the Canadian Federation of Agriculture received a phone call from the Department of Trade and Commerce, asking it to name one person to sit on an advisory council to the minister in regard to negotiations with the United States to achieve a free trade agreement. The Canadian Federation of Agriculture executive thought that I would be ideal for the role, but were concerned that members of the Ontario Federation of Agriculture might feel otherwise, since there was some concern that I might not be supportive of supply-management organizations.

I agreed to take on the task and met with the Ontario Federation of Agriculture in Peter Hannam's quonset. The next day I met over lunch

5 Ibid., 89–90.

with the Ontario Wheat Board. Both groups were satisfied with my intended approach, so I took a seat on the International Trade Advisory Council. At the onset, only Walter Kroeker, a potato grower from Manitoba, and I represented agriculture. Ray Malinowski from Yorkton was the only other person from Saskatchewan and thus, Saskatchewan and agriculture were each under-represented.

There were about 40 people on the committee, which was heavily represented by Ontario and Quebec, since that was where most businesses were located. Most of the committee members had very little understanding of agriculture and were not really interested in our industry. So, Walter Kroeker and I had our work cut out for us. Walter tended to be shy and had to be invited into a discussion, although once in, he was very good. I soon learned that I had to be assertive and talk about agriculture, whether anyone else did or not.

I defended the Canadian Wheat Board against attacks and tried to create an understanding for the need for marketing boards. Much to the surprise of many, I gave examples of where supply management was more efficient than so-called free enterprise. I ruffled a few feathers but also earned the respect of many around the table. Ron Osborne of London Life told Al McLeod that, "Turner sure has big balls!," since I wasn't afraid to challenge or to make my case.

We were fortunate to have an outstanding committee chairman, Walter Light of Northern Telecom. He was very fair and by no means a patsy for the government. The minister or his/her deputy rarely attended the meetings. When there was disagreement with what the minions were pushing, Light would go to the minister's office to set the matter straight. Walter had done a good job of growing Northern Telecom and thus had the aura of success on his side. Trevor Eyton was also a good chairman but not as effective as Light. It seemed that Trevor was more prone to pushing the bureaucratic line. John Crosbie from Newfoundland became minister and was also a no-nonsense individual. Light and I became friends and while he did not understand agriculture, he did respect what I was saying.

Few at that table even realized that the United States was subsidizing its farmers, the way it was supporting agriculture exports, nor the subsidies attached to the Mississippi River waterway. I was often furious with our Canadian civil servants because they refused to comment in our documentation on the unfair advantage the United States had by using a waterway that was built by public funds and maintained by the Army Corps of Engineers at no cost to the shippers of products or the users of the river.

There were several task forces created and I was asked to chair that on agriculture. Alan McLeod and Murray Bryck of the Saskatchewan Wheat Pool research division, with my consultation, put together an excellent report. My next seat neighbour, Jim Smith of Domtar, a believer in free enterprise, described the report as a wish list for the industry. Although it wasn't meant to be, I thanked him for his compliment. It bewildered me to understand that when you are establishing a negotiating position, that you would not outline your ultimate objectives within your own group.

To complement the International Trade Advisory Committee, smaller groups were organized—Sectoral Advisory Groups on International Trade (SAGIT). It was at this point that Bill Vaggs, president of the Manitoba Hog Producers Association, entered the picture. He was an articulate and effective spokesman for hog producers and, indeed, for agriculture. Agriculture was underrepresented on the ITAC, and I could have benefited from Vagg's presence earlier.

GENERAL AGREEMENT ON TRADE AND TARIFFS

I was asked to stay in an advisory role for these international negotiations. I did so and I expect because of my previous experience and the confidence that Walter Light had in me, I was included in the Canadian delegation to the opening of the General Agreement on Trade and Tariffs in Punta Del Este, Uruguay, in September 1986.

This was a great experience and grains got a lot of attention. Our delegation was sought out by each of the United States and European delegations who were seeking our support for their particular positions. We staked out our position of only agreeing if it would rattle the high subsidies paid by Europe and the United States. Each party was intent on blaming the other for the problem. This was just the opening session and, of course, nothing was achieved to reduce export and production subsidies. So, too, because of the massive supply management in Europe, the Canadian programs were safe.

There was a move by Canada and the United States to get agriculture fast-tracked on the GATT agenda, which meant early and special attention in negotiations. This was successfully opposed by other countries because they wanted their concerns to be front and centre. As well, Japan benefited from subsidies which kept grain prices low, and Europe—especially France—wanted to maintain the Common Agriculture Policy, which was centred on subsidies. Therefore, the fast track tactic failed.

We launched an enthusiastic thrust by holding meetings on a delegation-to-delegation basis to gain support for ending grain subsidies. We desperately wanted to get rid of the subsidies since we were sure Canada would prosper in an uninhibited market place. One session with the United States was particularly volatile, with the Americans threatening Canada, and by inference, Europe, that "If you don't like our subsidies, you haven't seen anything yet!" It was an inflammatory statement, but not the official United States position. We were forthright with the United States about the damage its subsidy programs were doing to Canada. The United States admitted that its real target was Europe, and indicated that it was sorry we were caught in the backwash. This was nice to hear but not particularly useful.

We held a reception, but the turnout was so large and the name tags so small that it was not possible to seek out those whom we wished to persuade to the Canadian point of view. The Japanese strongly rejected the fast track for agriculture because it wanted the focus to be on services and on tropical products. It was obvious at the GATT launch that Europe was at loggerheads with the United States and Canada on the subsidy question, and very much in conflict with Japan on trade between those two entities. Europe had produced a document entitled "Balance of Benefits," which would give Europe greater access to Japanese markets. The Japanese felt Europe was taking an aggressive position with them and said the document really stood for "Bash Oriental Bastards." This identified one of many frictions between participants in Punta Del Este. Russia applied for status at the talks but was turned down because of strong opposition from the United States.

It became apparent that there would be no progress in addressing the agriculture subsidy issue. We left Uruguay in support of a suggestion that an Eminent Person Group be established—one person of stature, selected from each of several concerned countries, to help deal with this vital question of subsidies.

I returned home to the usual busy period leading up to the delegates' annual meeting of the Saskatchewan Wheat Pool. I remained on the committee for a while longer and then one day received a letter simply stating that my appointment for the next year would not be renewed, without further explanation. The bluntness of this letter was offset by one from Prime Minister Mulroney expressing his thanks for my time on the ITAC.

CHAPTER 9
New Opportunities

FAREWELL TO THE SASKATCHEWAN WHEAT POOL

The 1980s were proving to be very exhausting. In addition to the heavy workload of the president, other activities were being added, such as the protracted Crow Rate exercise. This involved many hours of preparation and the need to attend high-pressure meetings at our membership level and at the Ottawa level. Additionally, as discussed earlier, the XCAN affair was threatening to explode. As president, this would involve a tremendous amount of my time in damage control. Finally, although the ITAC meetings were not numerous, they did require a lot of time and preparation.

As an example of the time commitment involved, in the two-month period starting on August 26, 1986, outside of my regular Pool duties, I had to meet with the Ontario Federation of Agriculture executive, with Ontario Wheat Board officials, with Ontario Corn Growers manager Terry Daynard, with Canadian Pacific Railways representatives in Winnipeg, with Saskatchewan premier Grant Devine, and with the Federal Trade Secretariat in Regina. I attended the GATT negotiations in Uruguay from September 12 to 20, attended the Prairie Pools Board meeting, and went to Edmonton to meet with Prime Minister Brian Mulroney and his staff and farm leaders on September 25. There were International

Trade Advisory Committee-related meetings on October 8, 21, and 26, and a deficiency payment meeting in Ottawa on October 16.

During this period of time there were numerous internal Saskatchewan Wheat Pool meetings in preparation for the annual meeting. As well, I held meetings in my sub-district, a District 16 delegates' meeting in North Battleford, and another deficiency payment meeting in Ottawa on November 5. During this period, the usual Pool activities took place, including annual meetings of subsidiary companies and long-service award banquets.

This was a very hectic period. There was scarcely time to put one function behind me before preparing for the next. To add to the demand on my time, Leo Kristjansen, president of the University of Saskatchewan, asked the Saskatchewan Wheat Pool Board to allow me to chair a capital campaign to raise money for a new College of Agriculture building. The board invited Leo to come to the delegates' annual meeting in November to repeat the request.

These additional duties took me away from District 16 and my responsibilities to my fellow delegates. They also kept me away from Sub-district 1 and my duties to the farmers in that area. At the sub-district level, where I had not lived for almost twenty years, I could no longer call all the committee men by their first names. These new, younger members did not know me and therefore my infrequent visits to the sub-district were not enough to establish their confidence. In 1983, I was challenged for the first time in 25 years. The contest was close, but I won by 37 votes. The challenger had not participated in Pool activities and was not actively farming; his brother handled the land. He ran on the platform that I had betrayed farmers by daring to negotiate the Crow Rate. Two years later I was again challenged, this time by an active farmer, but not one who was well-known in the sub-district. Once again the vote was close and the issue—the grain rate—was the same. The candidate's father told me that his son had been encouraged to run by Ted Strain, past president of the National Farmers' Union. He and one of my delegates actively campaigned against me. Unfortunately, this was typical of the National Farmers' Union philosophy of tearing down, rather than trying to build up.

I announced to my committees that I would not seek the nomination in 1987. I was wearing down and my duties kept piling up. Even though I had won each of my delegate elections, it was clear that I was not communicating effectively at the sub-district level. When asked if I thought the challenges to me were unfair, I responded: "Every member of the Saskatchewan Wheat Pool has the right to seek to be a delegate, that is a strength of our organization. I have accepted the privileges of the democratic system; therefore, I must accept the results, whatever they are, of that system."

My questioner asked: "How can you be so philosophical at a time when it is possible that your whole life could change?" My response was that I am totally committed to the democratic system. What did annoy me was outside interference, such as an anonymous letter from a box number in Saskatoon, mailed to the shareholders in my sub-district, which was full of innuendoes and which attributed things to me that were not true. The letter was mailed just before the election period and thus I had no chance to set the record straight.

During the board election in April 1986, I was warned that there were discussions being held to prevent my election as president. Although I was not challenged as president, the election resulted in Avery Sahl, who came into the meeting as second vice-president, being elected as first vice-president and Garf Stevenson elected as second vice-president—a role switch. This bewildered me because Garf was doing an outstanding job. What was going on? I asked a few of the directors to meet with me for breakfast the next morning. I needed to know why the switch, so I could understand our board dynamics. I included Buck Sanderson, who had warned me the previous day, and our two executive members, Ken Elder and Leroy Larsen. However, I gained virtually no insight into what was going on behind the scenes, and ever since, I have been bewildered about the antics of the board on that day when they switched Garf and Avery.

Buck had warned me that I was in jeopardy, although not why my position was threatened? Why did they not carry through and oust me? Why pick on Garf? He was doing an outstanding job. Avery was adequate but at times not very communicative. I don't recall one time when he came into my office unless invited to do so. Garf would drop in and we would, together, decide the best approach to a certain challenge.

Perhaps I had been in the position for too long, but there was never any dissatisfaction expressed to me by any of the directors, nor about Garf's performance. Perhaps I had rubbed some of the directors the wrong way because I was intolerant of what I perceived to be any lack of total commitment to the policies of the Saskatchewan Wheat Pool. In any event, Garf carried on without any loss of enthusiasm or dedication, which I view as a real testimony to the character of an outstanding man. He, along with Bill Strath of Manitoba Pool Elevators and Doug Livingstone, of Alberta Wheat Pool, were the lead group in policy development for Prairie Pools Inc., which had just set up an office in Ottawa and had appointed Javier Caceras, who had been in the research division of the Saskatchewan Wheat Pool, as its manager. I will discuss Prairie Pools, and my role therein, in the next chapter. Suffice it to say here that I was approached by Bill Strath of Manitoba Pool Elevators, Doug Livingstone and Garf Stevenson in the summer of 1986 to take over the direction of the new entity.

I agreed to give it a try if we could work out the essentials of responsibilities and a pay structure. Perhaps I had been in the delegate, director, and president positions at the Saskatchewan Wheat Pool for too long. In the subsequent discussions, I was described as domineering, although the commentator indicated that this was a good thing. Interestingly, I thought of myself as mild, but I did advance my convictions with some vigour. In giving my opinion, I felt that I was putting out a statement for discussion; others evidently felt that, from the way I presented a point of view, there was no room for a contrary response.

By December, we had worked out the details of my new role. I would be called executive director of Prairie Pools Inc. and would be located in an office in Regina. I committed to a minimum of two years and a maximum of five years, commencing March 16, 1987.

It was a traumatic experience to leave the Saskatchewan Wheat Pool some 29 years after first becoming a delegate. I would miss the close association of many people. I would be on the outside when important decisions were made, and I would miss the prestige that went with the title of president of the Saskatchewan Wheat Pool. However, I would continue to advance public policy issues, and from a broader base. I would be allowed to continue as co-chair of the University of Saskatchewan Agriculture Building Capital Campaign, as a director of Philom Bios, on the Conference Board of Canada, as president of the Great Lakes Waterways Association, and on ITAC, so I indeed did have a very good situation.

At our December board meeting, the board was informed that I would resign all of my positions effective January 19, 1987, just a few months short of my 60th birthday. I called a meeting of District 16 delegates for December 19 at 9 a.m., where they were informed of my change of role. I had a meeting in my sub-district at 12:30 the same day, where, following the mid-day meal, I advised members of all my committees about my decision and urged them to seriously consider who would be a good candidate for delegate.

There was a great flurry of media interviews and good wishes, but I still functioned until the board meeting in January. A District 16 meeting was held on January 15 and 16, at which Dennis Vanderhaegen was elected as director in time for him to attend the board meeting on the January 19. In the meantime, I attended a Saskatchewan Wheat Pool Board meeting on January 7, at which they determined the format of the January 19 meeting where they would elect a new president. And so, after many wonderful years, I departed the control structure of the Saskatchewan Wheat Pool.

At the election for a new president to replace me, Garf won out over Avery. It was not long before Avery decided to leave his elected roles in the Saskatchewan Wheat Pool. Garf did a commendable job for six years as president.

CHAPTER 10
Public Policy for Agriculture

PRAIRIE POOLS INC.

ollowing the start-up of the three prairie Wheat Pools, it soon became evident and apparent that working together, primarily on activities outside of their respective provinces, would provide economies of scale. Originally, the Pools set up a selling agency—the Canadian Co-operative Wheat Producers—and purchased grain from members, then sold it on world markets and returned the profit as a patronage payment. This worked well until the economic crash of 1929. In that year, the available international price was lower than the price each Pool had paid to acquire the grain, and the Pools suffered severe financial losses. The Canadian Co-operative Wheat Producers survived as a fully-registered company, but it was not used again for commercial purposes. The company did, however, serve as an entity to publish an annual commentary about the needs of agriculture. The board of the Canadian Co-operative Wheat Producers was comprised of the president and two others, usually the vice-presidents, of each parent organization. February was the usual time when the annual meeting was held. Most often the president of the Saskatchewan Wheat Pool was also the president of the Canadian Co-operative Wheat Producers. The secretary revolved between the Pools, but for years, Don Richmond of Manitoba Pool filled that position.

In 1984, at the annual meeting, it was decided that since the Pools were already working closely together on public policy advancement, the function should be formalized. The Canadian Co-operative Wheat Producers was not descriptive of this new role and did not provide an obvious identity. Therefore, Prairie Pools Inc. was formed and we increased and concentrated our efforts to influence programs and legislation relevant to agriculture for the benefit of farmers. Ottawa was to be our main focus and we sharpened our activities by way of sending someone there for periods of time. That individual's job was to liaise with relevant bureaucrats and with the Canadian Federation of Agriculture office. This approach increased our effectiveness by the strategic use of the Pool presidents and/or a selected Pool president. John Trew, a long-term employee of the Saskatchewan Wheat Pool, was very effective in this role.

On one occasion, he arranged a series of meetings for me with relevant federal cabinet ministers. One such meeting was with Jean Chrétien. My purpose was to talk about the need for financial support for farmers and also about consequences of the Crow Rate situation. John and I were received into his office at 5 p.m. sharp. Chrétien started the meeting by pointing to a large wall clock and said, "When that big hand gets to the four, the meeting is over!" I thought, "Hmmmm ... twenty minutes! I'd better get started!" I began by stating, "Mr. Chrétien, thank you for meeting with us. I am here to talk to you about farmers' net income and about the Crow Rate." He immediately said, "you don't have to tell me about the Crow Rate. My grandfather farmed south of Calgary; I know all about it!" I responded, "That is interesting, but there are things about the Crow Rate situation I would like..." Chrétien jumped in again with, "My grandfather farmed south of Calgary; I know all about the Crow Rate." This was delivered in a slightly raised voice level, and with his delightful French accent. I replied, also in a slightly raised voice level, "I don't give a goddamn if your grandfather came over on the Mayflower! I am here to talk about the needs of prairie farmers!" From then on, we had an excellent discussion and John and I exited his office with a handshake. Although there was no handshake when we came in, we exited his office as the big hand reached the twelve mark! We had an audience of one hour. The lesson here—don't let politicians push you around!

The three prairie Pools were increasingly cooperating in areas of mutual interest: XCAN Grain, Western Co-operative Fertilizers Ltd. (also included was Federated Co-op), CSP Foods with the Saskatchewan Wheat Pool and Manitoba Pool Elevators (Alberta Wheat Pool pulled out at the last minute), the Federal Grain purchase, Pacific Elevators, Western Pool Terminals, Pool Insurance, and Pool Agencies. I felt there was a good comfort level within the Pools, in both commercial operations and in the public policy area. More and more, we were going forward together to advance our jointly arrived-at positions. In July of each year, the full board of directors of each Pool, with senior management personnel, would assemble to discuss commercial operations and policy objectives. Not only did this allow us to consolidate and strengthen our positions, but also to develop camaraderie among the people from each organization. Virtually every director and management person brought his wife to this function. It did much to achieve cohesion among our organizations.

On one other occasion each year, the executive (president and vice-presidents) of each Pool would get together, usually in February, and would work out a joint policy and a strategy for advancing it. Not surprisingly, Ottawa was the target for most of our efforts. It became obvious that even though sending capable people to Ottawa was a productive exercise, having a permanent person there would be desirable and so we opened the Ottawa office of Prairie Pools Inc. in 1985, with

A meeting of the Prairie Pools, 1978. Ted is to the left of centre, with the mustache, dark tie, and dark-rimmed glasses.

Javier Caceras installed as executive director. The costs were shared on a pro-rata basis consistent with each Pool's grain handlings. A common split would be the Saskatchewan Wheat Pool 55 percent, Alberta Wheat Pool 30 percent, and Manitoba Pool Elevators 15 percent.

While I was president of the Saskatchewan Wheat Pool, I was also president of Prairie Pools Inc. After my departure, the presidency revolved on a two-year basis. Because the president of each Pool was so heavily involved in other matters, the first vice-presidents as a committee dealt with the strategy of policy advancement.

When I agreed to take on the management role at Prairie Pools Inc., it became necessary to reconfigure titles and responsibilities. Javier protested strongly that I should be named executive director of Prairie Pools Inc. because it would diminish his importance in Ottawa, yet there was no other title that so aptly described the job I was to do. Javier, on his own, did not carry a presence, so he had to rely on the prestige of Prairie Pools Inc. He reluctantly accepted the position of manager of Prairie Pools Inc., Ottawa office. Another stumbling point was that Javier protested the lines of authority. Even though he reported to me, his appointment had been made by the board of Prairie Pools Inc. and he wanted to report directly to the board. I compromised and agreed to a board appointment procedure, but he was accountable to me for his work. This was not the best arrangement. Javier felt his role had been diminished when, had he been patient, I felt I could have elevated his status.

I spent time in Ottawa each month and was well-known by the ministries with whom we dealt. When I felt it was needed, Javier and I would arrange for the appropriate person or people to attend in Ottawa. I spent time in Calgary and Winnipeg at board meetings of the Alberta Wheat Pool and Manitoba Pool Elevators, and also did some work in the country. Of course, I also reported to the annual meeting of each Pool.

We were in numerous discussions with cabinet ministers and were pressing for deficiency payments for grain, and also taking strong positions in defence of the Canadian Wheat Board. We were often at loggerheads with Charlie Mayer, minister for the Canadian Wheat Board. Charlie was pushing hard to establish variable freight rates for grain and we were resisting. I was gradually building a positive relationship with his office, and Mayer was becoming more receptive to Prairie Pools Inc. and the Pools in general.

Then, an article very critical of Charlie Mayer and the federal government appeared in the *Western Producer*. It was an interview with Caceras, saying that the government had co-opted the Wheat Pools, leading them to believe certain things would happen, and then went back on its word. To my dismay, not only was the article incorrect, but it essentially accused Mayer and others of dishonesty. Mayer phoned me, very angry, and I could see that all my careful efforts at liaison were going out the window. I told Mayer that the article did not represent the feeling of the Pools and that he should have his staff contact Javier for an explanation, and that I would tell Javier he had better accommodate the wishes of Mayer.

In addition to the inaccuracies, Caceras was a facilitator, not a spokesman. I was to speak on such matters unless there was an officer of the Pools available, and then I would defer to that person. If no one was available, then I was to comment. I pacified Mayer and phoned Bill Strath of Manitoba Pool Elevators and the president of Prairie Pools Inc. I told him that I thought we could correct the damage Javier had done. I got a strong rebuke; something to the effect that I was now kowtowing to Charlie Mayer, and was told to leave the situation alone. When Strath failed to support me, I knew I didn't want to continue in that role. Had we put in place the proper reporting structure, it would have been handled. Caceras had constantly cultivated Strath, and I believe, had undermined me.

The Pools were experiencing a financial downturn, and the prospect of closing the Regina office was appealing. So, after two full years, we closed the office and I continued on a fee-for-service basis for a short period. The lesson here was that I agreed to accept a less-than-sound situation and was made the scapegoat.

Caceras never trusted me as his supervisor because I had made it clear to him that he should not be involved in partisan politics. He was strongly inclined to the New Democratic Party (NDP) and spent much time with NDP members of Parliament. Such things do not go unnoticed, and it jeopardized his ability to gain trust from those in power. He was inclined to be furtive rather than forthright, a complete contradiction to my style.

THE CANADIAN FEDERATION OF AGRICULTURE

All during my tenure as a director of the Saskatchewan Wheat Pool and up to my final days with Prairie Pools, Canadian farm organizations and farmers were well-served by the Canadian Federation of Agriculture (CFA). The CFA was in tune with farmer's needs because it had direct lines to the farm level by a broadly-based (national, in-scope) board of directors. The leadership of the CFA was always outstanding. The board was elected and headed by such remarkable men as Charles Munro and Glen Flaten, and the staff side had outstanding individuals—Dave Kirk, whose understanding was complete, with tireless energy, and Bill Hamilton, who was a complete leader with great sensitivity and a manner that inspired confidence in all around him. This organization and leadership gave us effective entry into international forums.

THE INTERNATIONAL FEDERATION OF AGRICULTURE PRODUCERS

The International Federation of Agriculture Producers (IFAP) was a very effective organization with a global structure; the membership was comprised of a farm organization from each of the 30 member

With the Canadian delegation of the International Federation of Agricultural Producers in Paris in 1971. Mel is in the striped dress in the middle of the front row; Ted is standing behind her wearing the dark-rimmed glasses.

countries. IFAP met at 18-month intervals and the discussions sought to bring world attention to agriculture problems that were universal in nature.

At the conclusion of each gathering, a profound statement was issued that described the global situation of agriculture and identified the requirements for a healthy world agricultural industry. One such statement from a session in Austria, at a time when there was concern about adequate food supplies, was entitled, "Food Security—the Key Farmer Security." A strong message was delivered in the title alone.

IFAP did much work in helping developing countries to establish productive agriculture industries. Canadians Charles Munro and Glen Flaten were outstanding IFAP presidents. IFAP was a forum that encouraged individual farm organizations to be aware of the needs of others while shaping their national farm policy positions.

INTERNATIONAL FARM POLICY LEADERS

During my tenure as president of the Saskatchewan Wheat Pool, I had the good fortune to meet and to discuss public policy for agriculture at the world level with numerous skilled and dedicated "farm leaders." Three such men stand out in my assessment of their impact on world agriculture.

Gene Moos was the president of the National Association of Wheat Growers (NAWG) and farmed in Palouse County in the state of Washington. Gene and I had an easy rapport with similar objectives for farmers. Gene was intrigued by the Canadian Wheat Board and our marketing system. On several occasions, he invited me to speak at the annual meeting of the NAWG. There was tremendous interest by those in attendance, but the board of NAWG was not interested in suggesting any changes to the US marketing system, so there was never any follow-up to my remarks. It was interesting that at its annual meeting, every private grain company had elaborate displays and were working the crowd, and, I expect, contributing to offset the cost of the meeting. Gene spoke at the annual meeting of the Saskatchewan Wheat Pool and was supportive of our system, but was unable to persuade his board to even investigate something similar for the United States. Gene became an advisor to Democratic senator Tom Foley and moved to Washington, DC. That was where I had my last contact with him—a fine person and an articulate spokesman for farmers. Mel and I visited him at his farm and enjoyed a visit with him to nearby Coeur D'Alene, a very lovely summer resort.

Sir Leslie Price was chairman of the Australian Wheat Board. Leslie and I became very good friends and attended many of the same international meetings; our messages were very similar. As well, we had an informal alliance at the International Grain negotiations in Geneva. Les was highly respected around the world and carried his farmer representation duties proudly. He was a sought-after speaker at grain forums in many countries. I recall one occasion in Winnipeg, the day following the meeting, when we awoke to 18 inches of snow. Taxis were not operating and I learned that Les had checked out of the hotel. He told me later that a four-wheel drive truck had pulled up at the hotel and Les had hired the driver to take him to the airport. Upon arrival there, he sought out the airline that would take him the farthest south at the soonest time. He ended up in Texas, but at least he was away from the dreaded snow that he feared and had never experienced before. Mel and I visited Les and Lorna in their home in beautiful Toowoomba, where I saw first-hand the operation of the Queensland Wheat Growers Association. We were delighted when Les was knighted and he and his wife became Sir Leslie and Dame Lorna Price.

Following one of the negotiation sessions in Geneva, I received a phone call from Tony Dechant, president of the USA National Farmers Union. Tony wanted to know what position the American delegation had taken in the discussions. I expressed my disappointment that they had supported a position of a ceiling price that was equivalent to Canada's proposed floor price. His next question was, "Did the advisors to the American delegation also support that position?" I said, "Yes, they did," and that I had argued with them in that regard. I had gone to Geneva expecting that Canada and the United States would have very similar positions but this was not so. A man from Cargill (one of the advisors to the American delegation) tried to persuade me to support our proposed floor price as the ceiling so all pricing would be below what I felt was a disaster level for our farmers. The philosophy of this approach was that it would destroy production in most countries, and soon the United States and Canada would have a monopoly of the market. I vigorously rejected such an approach. Dechant, as NFU president, called all the advisors to a meeting and exposed their position; most of them were replaced. The United States NFU position on most things was similar to ours. I spoke at its annual meeting several times and Dechant at the annual meeting of the Saskatchewan Wheat Pool once. Tony was also very active in the International Federation of Agriculture; it was great to have an articulate spokesman whose approach was similar to ours.

These three men—Moos, Price and Dechant—were dedicated farm leaders who each set a fine example of conduct in their personal lives and were my good friends for a period of time.

THE CHAMBER OF MARITIME COMMERCE AND THE
GREAT LAKES WATERWAYS DEVELOPMENT ASSOCIATION

The three Pools, through the joint agency of the Canadian Co-operative Wheat Producers Limited, were long-time supporters of the Great Lakes Waterways Development Association (GLWDA), whose main purpose was to promote this marvellous inland waterway. The Pools, because of their massive terminals at Thunder Bay, had a very real commercial interest in the effective functioning of this section of the St. Lawrence Seaway and were the largest financial contributors of any member of the Great Lakes Waterways Development Association. However, the Pools did not take a leadership role in the association structure. There was a large membership consisting of virtually all the grain companies, shipping companies, steel manufacturing companies, coal companies, individual ports, Ontario Hydro, and any other commercial entity that utilized the Seaway.

When I became executive director of Prairie Pools Inc., I was asked, and the board of Prairie Pools Inc. agreed, that I should become chair of the board of the GLWDA. This I did, and I worked closely with Norman Hall, president of the GLWDA, and with his assistant Peter Smith. They were based in Ottawa, and it was in the association offices where we held most of our meetings.

I was impressed with the quality and diversity of directors and their dedication to the role of director. Shortly after I began my tenure as chair, we changed the name of Great Lakes Waterways Development Association to that of the Chamber of Maritime Commerce (CMC). This, too, signalled a fresh and vigorous approach to our liaison and lobbying efforts. We were constantly after the St. Lawrence Seaway Authority, a federal government agency, to upgrade the locks on the waterway and to lengthen the shipping season. Normally, the period when ships could sail the full waterway was from early April to December 20, depending on ice conditions. Ships were generally frozen in for the winter months wherever they were located on December 20, and could not move again until early to mid-April. Locks were often drained on December 20, which allowed for maintenance and upgrading during the three-month non-operating period.

I really enjoyed this opportunity to serve. Each member of the Chamber could send a delegate to the annual meeting and in the case where a member already had a director, another representative was allowed—in our case, it was the chairperson of Prairie Pools Inc. The delegate list for the Great Lakes Waterways Development in 1988 was 183, representing shippers, ship operators, communities, government agencies, grain companies, coal companies, steel companies, and port authorities.

I spoke at a Great Lakes Mayors' Conference in May 1989 and identified the value of the Seaway system. Also, I talked about the shifting dynamics in grain movement as new markets opened up in Asia, and thus increased shipments from Pacific ports. I cautioned that the changes to the Western Grain Transportation Act that every other speaker was calling for, might shift even more grain westward—the exact opposite of what they expected.

Friction started when Hall and Smith sought to have the Chamber of Maritime Commerce state publicly its support for producer payments related to the Crow's Nest Pass as outlined in the Gilson Report. I refused to allow this because the Chamber of Maritime Commerce was not designed to be a policy organization and because it would not benefit the Chamber of Maritime Commerce. In fact, as I tried to explain to Hall and Smith, such payments would likely encourage the movement of grain to the West Coast and be detrimental to the St. Lawrence Seaway. However, they instead chose to listen to the Thunder Bay Harbour Association and members of the grain trade, who were in competition with the Pools. This conflict started to open a chasm between me as chair and Hall as president.

I spent a huge amount of time on the affairs of the Chamber of Maritime Commerce, attending large and small meetings. I was invited to attend a meeting of the Coalition on Grain Movement 1988, in Montreal on July 11, 1988. Everyone else in the meeting was advocating the demise of the Western Grain Transportation Act and of the

Canadian Wheat Board. I stood my ground and challenged them on every remark about these two entities. Once again, I warned them that they should be careful about what they wished for, as it might come back to haunt them.

There were numerous meetings and seminars that were loaded with presenters who advocated removal of the Western Grain Transportation Act. Often, I was the only person on the other side of the question, and I constantly challenged their assertions. I believe, on most occasions, I more than held my own. Cargill seemed determined to get rid of the Western Grain Transportation Act and to have payments made to producers, no doubt to allow them free reign in the country and to augment the handlings of their transfer terminal in Baie Comeau. I am afraid I acted more as an alley fighter than as a diplomat. I generally wrote letters to those who participated in the meetings, raising concerns about their remarks and, in reality, repeating the challenge I had expressed in the meetings. I lift a paragraph from a letter I wrote to the Honourable Ed Fulton, minister of transportation for Ontario. This paragraph is very similar to my remarks in other letters:

My point of warning in the meeting on May 19 was that strong supporters of the Seaway system must be very careful in the kind of pressure they bring to bear on the Western Grain Transportation Act since it is quite possible that significant change or elimination of the Western Grain Transportation Act could hurt the Seaway by way of reduced volumes.

Further in the letter I stated: "So, a word of caution: under a 'pay the producer' regime—Pacific Coast ports may have an even greater advantage over Thunder Bay than exists at present."

One other incident caused some of the other directors from Ontario to be less than supportive of me. I was chairing a public meeting in central Ontario to talk about the Seaway situation. A politician—a member of Parliament for the area, who was a member of the governing party—came and made relevant but challengeable remarks. I knew that his remarks would prompt questions. At the end of his speech, he said he had to leave right away for another function, and was gone out of the room—like "right now"! When I got back to the podium, I said, "Don't you just love hit and run politicians? They come in, lay it on you, and before you can respond, they're gone!" Apparently several of the Chamber of Maritime Commerce members did not like my jibe; they did not understand the Saskatchewan way of speaking forthrightly.

During the ensuing months, Hall continued to push the producer payment issue. With the support of the majority of the board, we amended the bylaws of the Chamber of Maritime Commerce. One clause stated that the chairman could only serve for two years, which was passed at a board meeting. I supported the change, expecting it to go into effect at the next annual meeting, and never thought it would be retroactive. The next annual meeting was in early May 1990, and the day before I was to leave for Ottawa, my sister Wyn died, so I did not attend the meeting.

Bill Strath, as president of Prairie Pools Inc., told me that when he got to the annual meeting, it was already arranged for a mover and a seconder to replace me with Ted Allen of United Grain Growers, who had only attended one Chamber of Maritime Commerce meeting. It was all carefully orchestrated ahead of time. Thus, my last service to the Chamber of Maritime Commerce was a few months prior to the annual meeting. I would have had more respect for Hall and Smith, had they the courage to contact me and to tell me what was happening. The upshot of this was that Prairie Pools Inc. let its membership lapse, and thus the Chamber of Maritime Commerce lost its largest contributor. Thereafter, the Chamber of Maritime Commerce became a non-entity as far as Prairie Pools Inc. was concerned.

Following being ousted from the position of chair of the board, I sent letters to several people. In one to Alex Graham, the chair of Prairie Pools, I explained that it would be costly to Prairie Pools Inc. for me to continue as a director of the Chamber of Maritime Commerce—a cost that could be avoided. As chair, the Chamber of Maritime Commerce paid my travel and related costs; Prairie Pools Inc. paid me a $600 per diem. If I were to continue as a director, Prairie Pools Inc. would have been faced with the travel and related costs plus the per diem. I suggested that Graham himself should attend, and then the only costs to Prairie Pools Inc. would be travel related. I meant it as a letter of termination from the board of the Chamber of Maritime Commerce, and Prairie Pools Inc. received it in that light, so after several years on the board, my participation ceased. After being replaced as chairman by a carefully orchestrated procedure, I received a letter dated May 18, 1990, from John Leitch, vice-chair of the Chamber of Maritime Commerce thanking me for my "valuable service" to the company, saying that they had intended to talk to me prior to the meeting and that the change was dictated by the new by-laws. I responded in a curt fashion in a letter dated June 1, making the following points:

1. That, would someone who had only attended one meeting be a suitable new chairman?

2. That I felt we had made considerable progress for the Chamber of Maritime Commerce under my chairmanship.

3. That, when was I to be told? Five minutes before the meeting?

4. That I had never heard of bylaws being retroactive.

5. That when John and John Hood had met with me to discuss my possible chairmanship, I felt I was dealing with two gentlemen with whom my first impressions were those of openness and honesty.

6. I raised the advisability of Norman Hall being president of the Chamber of Maritime Commerce and also of the Canadian Shipowners Association.

In conclusion, I once again warned of possible unfavourable consequences to the Seaway system if "pay the producer" was introduced in place of "pay the railways." Further, I stated, "I will continue the debate no further. I have received a clear message. It was kind of you to write. Yours sincerely." Thus ended my last act related to the Chamber of Maritime Commerce.

Bill Strath recommended to Prairie Pools Inc. that they withdraw their membership in the Chamber of Maritime Commerce, and such was done.

CHAPTER 11
Specific Policy Concerns

THE CROW RATE

In the late 1960s, Otto Lang burst upon the political scene as a Liberal member of Parliament for Humboldt. A former dean of law at the University of Saskatchewan, he was a brilliant man. Because of his intellect, Prime Minister Trudeau appointed Lang to his cabinet, and being from Saskatchewan, it was natural that Lang would assume agricultural-related responsibilities. My first exposure to him was during the wheat glut of the late 1960s. He designed and promoted, with the help of industry leaders, the Lower Inventories for Tomorrow (LIFT) program.

I was the only leader to stay the course in the application of this unpopular but necessary program. There was no doubt that Lang appreciated the support and we had a good relationship until he started to raise questions—first, about the usefulness of the Canadian Wheat Board and second, when he began to suggest that the Crow Rate was detrimental to agriculture.[1]

---

1 The Crow Rate was first instituted to guarantee a low freight rate for moving wheat from the prairie provinces to export position. In 1925, the rate was extended to other grains and was set at one-half a cent per ton per mile. Further to that, this was guaranteed in perpetuity.

In October 1974, at a meeting of the Canada Grains Council, Otto Lang raised the question of changing the Crow Rate and made suggestions of how it could be replaced. I denounced his remarks as being unnecessarily public and belligerent, and stated that he should have met with farm leaders to discuss freight problems.

At the Pool annual meeting in 1974, in the opening address to delegates, I noted that "maintaining favourable freight rates for grain" would be one of the major challenges facing farm organizations in the years ahead. I stated that there was a campaign underway to gain public support for an end to the Crow's Nest Pass Agreement. I challenged the railways, who claimed to be losing money hauling grain, to put their books on the table to allow a complete examination of the financial situation.

I quoted a position paper that the three Wheat Pools had developed which stated that "The Crow's Nest Pass rates for grain should be maintained." The paper then went on to identify ways to help offset railway costs, such as producers and government purchasing hopper cars, and government assisting with the repair of boxcars. Finally, I challenged the contention that the Crow Rate discriminated against the livestock industry.

Lang maintained that farmers would receive the same benefit as the Crow Rate offered, but it would be paid in a different manner; his first choice was as acreage payment. He did not seem to understand that farmers regarded the Crow Rate legislation as inviolable and were inclined to mistrust anyone who proposed changes to it. So, we were at loggerheads and the gap would only widen. Unfortunately, anyone who dared oppose what Lang suggested was treated as the enemy.

It was reported to me that Lang said he would organize his own farm group to counter the positions of the Saskatchewan Wheat Pool and other similar-minded farm organizations. This he did, aided by Ernie McWilliams of the Winnipeg Grain Exchange; thus, the birth of the Palliser Wheat Growers (PWG). Lang, with Palliser support, sought to make it easier to remove the Crow Rate and to get rid of the Canadian Wheat Board.

PALLISER WHEAT GROWERS

At the outset, the PWG seemed to be genuine in its attempts to promote wheat and to solve some problems in the grain industry. It held some of its formation meetings in the boardroom on the second floor of the Saskatchewan Wheat Pool building, so the Saskatchewan Wheat Pool was actually encouraging this new organization.

Walter Nelson, a successful businessman and farmer from Avonlea, was the first president. Soon after the PWG was established it became very critical of the Saskatchewan Wheat Pool and the Canadian Wheat Board. However, I was always able to hold a positive discussion with Nelson.

There was a grain workers' strike at Vancouver and the PWG berated the Saskatchewan Wheat Pool, saying we should just resolve the strike and get the grain moving again, and that we should forego all storage charges until a settlement was reached. Walter asked me to meet him for lunch, at which time I told him of the Grain Workers' Union demands. He was sympathetic to our position, realizing he would not want to pay such increases under similar circumstances. We heard little more about the situation from the PWG.

Nelson was generally fair but many of his colleagues were less so, portraying everything that went wrong as the fault of the Saskatchewan Wheat Pool or the Canadian Wheat Board. Our directors and delegates reacted angrily to what were mostly false accusations. There was no doubt that the PWG was firing the bullets that the Grain Exchange, many private grain companies, and even Lang's staff were making.

I spoke at the PWG's annual meeting on two occasions and was well-received. This showed me that only some of the PWG members were being so critical of the Saskatchewan Wheat Pool. The PWG seemed to attract those who were furthest to the right of the political spectrum.

There was initially a surge of interest in this commodity organization and its membership grew rapidly. However, it declined when people realized that the PWG was more inclined to destroy than build. Had Nelson kept his most outspoken members in check, they could have been very effective and useful in building the wheat industry.

Individuals and groups in Alberta, and indeed the Alberta government, found the disruptiveness of the PWG to be attractive, and so its declining membership in Saskatchewan was, in part, replaced by new members from Alberta.

However, governments took the PWG into account, especially when trying to discredit the Pools. I often wonder what their farmer membership would have said if they had realized that, as part of the

Prairie Farm Commodity Coalition, they were advocating the Crow Rate deficit be paid to producers. Had the federal government got its way, payments would have been on an acreage basis and there would have been a huge transfer of money from grain producers to others as well as a huge transfer of money from Saskatchewan to Alberta. What a shame that a potentially good organization did so little to promote its own crop!

Action by the PWG, Otto Lang's position on the Crow Rate, and his attack on the Canadian Wheat Board were a red flag to Saskatchewan Wheat Pool members and the battle lines were drawn. This stubborn attitude by Lang, coupled with his disregard for farmers, prompted Davey Steuart, deputy premier of Saskatchewan, to remark that "Otto's head is so full of brains there is no room for common sense." Lang's actions at the federal level created problems for the Liberals in Saskatchewan. Lang engaged Carl Snavely, a highly regarded American transportation economist to investigate the cost of moving grain, and he determined that the railways were indeed losing money hauling grain under the Crow Rate. The Saskatchewan Wheat Pool participated actively in the Snavely Commission and was convinced of the study's accuracy.

However, Lang was successful in encouraging organized, critical groups clamouring for changes to the Canadian Wheat Board and to the statutory grain rate. The federal government and the railways were advocating branch line abandonment and elevator consolidation. The Saskatchewan Wheat Pool resisted this thrust because we felt the best interests of farmers would be served by farmer-owned companies making the major decisions in this regard. The Saskatchewan Wheat Pool, for instance, presented plans to the membership that would reduce the number of Pool delivery points from 1200 to 650 over five or six years, and then eventually to 400 on a permanent basis.

When Lang lost his seat and left the government following the defeat of the Liberals in the 1979 federal election, the pressure from Ottawa lessened briefly. However, the PWG kept up the fight, abetted by the government of Alberta and livestock groups. The PWG and several other commodity groups merged to become the Prairie Farm Commodity Coalition (PFCC). Lang's acreage payment suggestion excited many landowners who were not primarily grain producers.

ENTER JEAN LUC PÉPIN

The Progressive Conservatives under Joe Clark, who had defeated the federal Liberals in 1979, in turn lost power in 1980. The Liberals under Pierre Trudeau were back in office, but without Otto Lang. Jean Luc Pépin was appointed as the minister of transportation and a new assault on the Crow Rate ensued. This was encouraged by many Western Canada organizations, including the Western Agriculture Conference, an assembly of farm organizations from the area west of the Manitoba/ Ontario border, and who were within the structure of the Canadian Federation of Agriculture. There was a growing concern by shippers of all commodities in the west about railway capacity in the near future.

Grain, potash, and coal producers all foresaw dramatic increases in the volume of freight required. The railways were claiming that they were losing so much money moving grain that they were unable to make the required changes to upgrade their systems. A logical question to ask was, "Can anything really exist in perpetuity?"

In February 1980, the Western Agriculture Conference met in Regina and tabled a resolution calling for the federal government to review the statutory freight rate with a view to increasing railway capacity. Bill Marshall, second vice-president of the Saskatchewan Wheat Pool and an executive member of the Western Agriculture Conference, along with Glen McGlaughlin of our research division, came into my office and asked for my permission for them to support the Crow revision resolution that would be dealt with at the conference, that day. After a lengthy discussion and with a remarkable lack of enthusiasm, I agreed. It was essential that railway capacity be expanded and for the railroads to become enthusiastic in doing their job.

I knew that to even suggest tinkering with the Crow Rate would cause great trauma within the membership of the Saskatchewan Wheat Pool. This was an understatement, and my new position would haunt me for the rest of my days as president. However, it was the right decision and the resolution passed! We were committed and it was then necessary to prepare a specific position for Saskatchewan Wheat Pool delegates to consider. This we did, with input from members at meetings, leading up to our annual meeting in November 1980.

In my address to the 1980 Saskatchewan Wheat Pool delegates' meeting, I spoke extensively on the Crow issue, and outlined the Saskatchewan Wheat Pool position: (a) the railways must be adequately compensated for moving grain; (b) producers should continue to pay the Crow rates; (c) revenue shortfalls should be paid by the federal government directly to the railways; and (d) subsidies should be conditional on railway performance. I also spoke against concepts such as variable rates and acreage payments to producers. Further, I made the case that change was coming, and that the Saskatchewan Wheat Pool had to be at the table when negotiations got underway, in order to protect the interests of its members. The November 10 issue of the *Leader-Post* reported that, "Turner was displaying considerable leadership, and not without some political risk to himself." There was a protracted and vigorous debate, and a very good resolution was adopted by a recorded vote of 122 in favour to 22 opposed. Resolutions at the Saskatchewan Wheat Pool annual meetings required a two-thirds majority, so this endorsement was very strong. The minutes of the 1980 Annual Meeting of delegates indicated that the following resolution was passed:

That the Saskatchewan Wheat Pool support the Western Agricultural Conference transportation policy proposal, including the following:

1.   Grain rates to be statutory;

2.   Guaranteed performance by the railroads;

3.   Consideration to freight rates on processed products;

4.   Rate structure to be distance-related;

5.   Compensatory rate shortfalls to be paid directly to the railroads;

but within this policy, Saskatchewan Wheat Pool press for the retention of the present statutory Crow Rate for the producer.

This resolution sparked lively and sometimes very critical comments at member meetings. Generally, however, if a reasonable discussion was held, the members agreed with the stated policy. Producers knew that an efficient grain transportation system was essential, albeit at a higher cost. They also were not prepared to hand to the railways the right to configure the system by way of variable rates, nor were they prepared to see recompense for higher rates handed to others who did not even ship grain, as would be the case in payments to producers on an acreage basis.

The following year was a hectic one, buffeted on one side by numerous member meetings and by district delegate meetings, and on the other side by keeping government officials from trying to change the grain rate, while ignoring the five conditions that were part of the main resolution. Pépin and company worked full-time at trying to bring the federal government onside (there were cabinet ministers who opposed any change), and, at the same time, tried to manipulate producer organizations into a supportive group for their purposes.

November 1981 and the 57th Annual Meeting of the Saskatchewan Wheat Pool arrived. There was little doubt that the major topic for debate would be our grain rate position. True to expectations, the debate dealt with each of the five points separately, and three new points were added to become an eight-point policy. The essence of our previous policy was unchanged, but the level of support dropped from 122 to 99 in favour, and 45 opposed, compared to 22 for the previous policy. This, however, was still strong endorsement and allowed us to advance the policy with vigour.

A REVIEW PANEL

In order to deal with the areas of disagreement, Pépin and Arthur Kroeger, Pépin's deputy minister, decided to set up a broadly based agricultural industry group and thus move the decision away from grain producers only. The government forces moved swiftly with a concerted effort to add organizations that would oppose the conditions put forth by the Saskatchewan Wheat Pool and endorsed by the Manitoba Pool Elevators and Alberta Wheat Pool. Kroeger organized a very formidable government team. They decided that the method of change should be decided by a group of Western Canadian farmer organizations. Also included would be the CNR and the CPR, as well they should have been, along with each of the Pools, the United Grain Growers, the Manitoba Farm Bureau, livestock groups from Alberta, and the Prairie Farm Commodity Coalition—which was pro-change and willing to do the government's bidding on the issue.

CHAPTER 11

The Saskatchewan Wheat Pool opposed the inclusion of livestock groups because it was a grain rate discussion. However, Kroeger and company were prepared to go to any length to dilute the influence of the Pools, and especially of the Saskatchewan Wheat Pool. They constantly attributed incorrect motives to us for the positions we took, such as suggesting that we took certain positions in order to protect our "antiquated elevator system" when, in effect, our elevator system was the most up-to-date of all and we had done more to rationalize our system than any other company.

DR. CLAY GILSON, CHAIR OF CROW RATE COMMITTEE
Kroeger proposed Mac Runciman as the chair for the discussions. The Pools opposed him because he was deeply biased against all the things we thought important, and he leaned heavily toward the railways. Kroeger said the Pools opposed Runciman because he was market-oriented, implying that the Pools were not.

The truth is that we beat the United Grain Growers and the other companies at every turn and they achieved very little original in grain matters. After all, Runciman had withdrawn the United Grain Growers from XCAN because he couldn't keep up with the market-oriented Pools. Interestingly, I asked Kroeger some years later what he meant by "market-oriented." He was at a loss to explain, because, in reality, a deputy minister does not market any tangible product.

Finally, the federal team proposed that Clay Gilson, a University of Manitoba professor and an internationally-respected agricultural economist, be chairman. We readily agreed and he was appointed. The talks were scheduled to begin in early April, so we called all of our delegates to a meeting in Saskatoon on March 11, 1982, at the Holiday Inn. The purpose of the meeting was to advise the delegates of the upcoming negotiating session and to ask for further input into advancing our policy position. The meeting started with an update by me, as president. A motion to affirm our policy position was tabled. Then, a draft of a document to be used at the Gilson sessions was presented and, after considerable discussion, adopted.

We received a request from a group of 500 "Wheat Pool members" assembled in the Saskatoon auditorium for the delegates to meet with them to hear a statement concerning the Saskatchewan Wheat Pool's transportation policy. The meeting discussed the request and granted the president a half-hour recess to attend the "Wheat Pool members'" meeting.

Accompanied by Ian Bickle, I spoke at the meeting, answered questions and received their statement. I believe they were so surprised I came alone that some of them gave me a standing ovation when I was leaving. In the final analysis, the delegates gave good advice to those of us who would be attending the Gilson negotiations, and were quite supportive of our role.

The negotiations carried on periodically for several months, and we were able to achieve most of our goals. Even some of our competitors in the grain business were supportive. The railways, especially the CPR, on many occasions, seemed to lean to our position. The livestock representative, Chris Mills, was quite disruptive, as was Lorne Parker of the Manitoba Farm Business Group, who had virtually no constituents. In the final position, we were able to gain for the railways $650 million per year as a subsidy to offset their losses from the Crow Rate.

Gilson recommended that in 1983–84, the cost increase would be shared equally by government and producers up to a maximum of 3 percent for producers. For 1985–86, producers would pay the first 3 percentage points of cost increases and share the next 3 percent equally with government, with a maximum of 4.5 percent, with remaining costs borne by the government.

It was during these negotiations that I became good friends with Russ Allison, a senior vice-president with the CPR who had sympathy for the potential financial burden on farmers. I always felt that without outside interference, the two of us could have solved the situation to the satisfaction of the railways and producers. The discussions were not just between the two most affected parties, the railways and grain producers, which would seem logical. The inclusion of livestock and some other groups that stood to benefit without any risk of higher costs, made it necessary for extreme positions to be taken by those representing the grain shippers' interests. This newly fashioned system functioned well; however, organizations and producers succumbed to the carrot of a Crow buy-out. With it went any continuing vigilance on the railways and the money received by producers was soon gone.

The Crow process was a complex undertaking that would have gone better had the Pools not been outnumbered by other organizations, many of which had little if any contact with, or accountability to, producers, whereas we were in constant communication with our membership. The Palliser Wheat Growers did have a membership base, but they were kept in business by the Alberta government and their numbers across the prairies paled in comparison to the number of active farmers the Pools had. We were part of a skilful government manipulation whereby the majority (farmers) were subjugated to the minority. Our positions were always misrepresented and our motives were deliberately misconstrued by some organizations, including the government. We were attacked by some farmers even though our position was arrived at through a time-tested democratic process. In retrospect, we were on a different level than the federal government bureaucrats. To them, it was a game to be won and a chance to collect some "brownie points" from the minister. To us, it was a crucial struggle to get the best possible result for our farmer members.

A big disappointment was my failure to persuade Ted Strain, president of the National Farmers' Union, to attend the Gilson sessions. He would have been a huge help to us, but rather than get involved, it was easier and less risky to just criticize. We would have done even better with his presence. Throughout this whole exercise, it became clear that Pépin, Kroeger, and their assistants felt they knew better than anyone else, and they became increasingly difficult to work with.

FOLLOW-UP ACTIVITIES
The Saskatchewan Wheat Pool was very opposed to any payments to producers because they would look like a farm subsidy rather than a transportation subsidy. After the Gilson Report was issued, that became the main focus of discussions. As well, the Pool was very opposed to any opportunity for railways to implement variable freight rates. Some government officials implied our position was to protect Saskatchewan Wheat Pool's "over-built system of wooden elevators," but I said it was to prevent giving the railways permission to configure the grain collection system as they chose. Finally, the Pool was opposed to the recommendation that producers should pay the full cost of freight above the 30.4 million tonne level.

There could be no doubt that the Saskatchewan Wheat Pool was under serious pressure from its members. I spoke to a meeting of 300 farmers in North Battleford in early 1983. They had assembled from every corner of the province; many I had never seen before. This particular meeting was chaired by delegate Dennis Vanderhaegen. The group in North Battleford called themselves "Concerned Pool Members," and they rejected our position. At the meeting, I was accused of "selling out." The meeting also "demanded changes in the election of the Pool President." I believe that I was successful in putting forth our position, however, since the audience gave me a hearty round of applause. Moreover, many knew me quite well and were upset by the unfair accusations.

At the Saskatchewan Wheat Pool delegates' meeting in November 1982, one of the resolutions from the country called on me to resign. The resolution was defeated unanimously, and was followed by a standing ovation in support of me. Bob Philips wrote in the *Western Producer*:

> This was the second standing ovation for the doughty fighter who has been subject to an unparalleled string of abuse from across the country and the countryside for his tough stand on behalf of Pool members. The first came after Ted's opening address, which he concluded with, "We have had a trying year. We have agreed and we have disagreed behind these doors. It will no doubt be the same in the next few days. But you can be sure that if we are going to win this battle, our voice to the outside and to government must be one and the same. That is our challenge."

In February 1983, Pépin called a meeting in Winnipeg to announce government plans to change grain freight rates. I boycotted the meeting because the proposals needed improvement and several items were inadequately addressed. I said that "the government proposals do a good job of recognizing the needs of railways and other shippers, but do not recognize the economic situation faced by farmers." The Saskatchewan Wheat Pool issued a statement in February 1983 which concluded: "Mr. Pépin has delivered a cruel blow to western agriculture. His proposals are unfair."

In April 1983, I led a delegation to Ottawa made up of 120 farmer-members, plus provincial and municipal political leaders, to present a petition with 110,000 signatures. "There was ample evidence that many in the West feared what payments to producers and variable rates could do to their communities, and the Pools' opposition reflected not only its own commercial interests but also that of its rural members."

At a private breakfast with cabinet ministers Pépin, Hazen Argue, and Eugene Whelan and again later, at a public luncheon with 20 members of Parliament and senators, I said, "It is hard to imagine a plan that could be more inflammatory to the people of Saskatchewan." At the next annual meeting of delegates, I said that "It was one of my proudest moments as president of this organization to lead a delegation that conducted itself in such a fine manner to totally impress the elected and appointed representatives of the federal government."

At the November 1983 annual meeting of delegates, I spoke about the events of the past year, including all the work that had been done on the Crow Rate issue. I also spoke of the toll the acrimonious debates had taken on me and on the Saskatchewan Wheat Pool:

> The transportation issue has served to split the farming community. It has been a devastating experience for farm organizations. At times, I have been the focal point for attack by many individuals—attacked for presenting and representing the policy positions of the Saskatchewan Wheat Pool. These attacks have come from people who felt our policy position was too generous. They have come, as well, from those who felt that it had not gone nearly far enough to meet their desires. I am pleased to say that, for the most part, the conflict has resulted from an honest difference of philosophy and approach to topics. However, in some cases, the attacks have been personal. Many others have experienced a very similar situation. Our goal must now be to again establish meaningful communication within the farm community and within farm organizations. To do so, we must overlook the personal hurt that we felt on many occasions, and this is not easy to do.

The method of payment controversy continued with the government desperately wanting to achieve payments to producers. This did not happen under either a Conservative or Liberal government. The budget of 1995 eliminated the annual Crow benefit subsidy of $650 million entirely, and made a one-time payment of $1.6 billion to producers. Thus the 12-year debate ended.

To close this section, I borrow from Gary Fairbairn's book, *From Prairie Roots* (1984), because his concluding remarks on the Crow represent exactly the situation that existed, and it describes the vacillating emotions that were at play:

> The fundamental question of whether to change Crow rates had, in effect, been decided in 1980 and confirmed in 1981 when Pool delegates supported their directors in deciding not to go to war for the Crow. It was an historic decision, even though it was implemented amid confusion and bitterness. The embarrassing vacillations and contradictions in stated Pool policy were regrettable, but they were also signs of healthy tension between two essential elements in a strong democratic organization: the duty to provide leadership and the responsibility to stay in tune with members. Pool leaders could not march in disciplined, perfectly co-ordinated fashion into a brave new post-Crow world, because their members were not yet ready to discard that 85-year-old bird. At the same time, Pool leaders could not blindly say "no" to change, because they were painfully aware of the need to bring more revenue into the transportation system so that future generations of farmers would be able to get their grain to market. It was a tough, thankless task. It also showed the potential benefits by being both a strong commercial organization and a democratic policy-making association. If the Pool had been only a policy organization, it would have had far less incentive to take a responsible approach to modernizing the grain transportation system. If it had been only a commercial organization, it would have had less reason to fight for key safeguards for farmers and its recommendations would have been less trusted by farmers. Despite all the frustrations and trauma involved, the 1980–82 revision of freight rate policy also reconfirmed two basic facts: that Saskatchewan Wheat Pool's views on agricultural issues were still important to farmers, and that the Pool continued to reflect the hopes and fears of rural Saskatchewan.[2]

2 Gary Fairbairn, *From Prairie Roots: The Remarkable Story of Saskatchewan Wheat Pool.* (Saskatoon: Western Producer Prairie Books, 1984), 231–32.

FARMER INCOME

The 1980s were a turbulent decade for farmers and for those who represented their interests. Undesirable things, each of considerable scope and of significant magnitude, came together within a span of a few years. Trade agreement attempts, the Crow Rate confrontation, and a shocking lack of revenue created turmoil that rocked the farming industry. I have covered freight rates and trade attempts in other parts of this book, so I now turn my attention to net farm income.

Periodically, over the years, farmers were faced with what came to be known as "the cost-price squeeze," where revenues failed to cover expenses. The 1980s were a decade where trade, freight rates, and grain prices were intrinsically linked. Increases in freight rates impacted net earnings, and export subsidies by other countries reduced grain prices with the same impact on farmers' take-home pay. Export subsidies by the United States and the European Economic Community reduced Canada's market share, which translated into lower gross revenues.

The period 1974 to 1976 was perhaps the most remunerative time ever for grain producers, as markets were large and prices excellent. Many farmers were enjoying unprecedented high net incomes. This sparked an epidemic of spending on all farm-related consumer goods, and farm residences and other buildings were either upgraded or built anew. Holidays were taken, often for the first time. It seemed like never again would farmers be short of money.

By the time 1980 arrived, farm machinery prices had doubled or even tripled. Interest rates had risen to usury levels and farm input costs had increased dramatically. Then, the price of grain declined to the levels prior to 1974, and net farm income disappeared completely. Interest rates on loans rose to 20–22 percent, and many farmers found it impossible to service their loans. Lending agencies were forced to foreclose on many loans and they were accumulating more land than they could handle.

A former Maymont neighbour, Murray Gray, who ran a balanced operation with livestock to complement his grain production, told me that because the income was so good, he decided to upgrade all his farm machinery. In order to do so, he took out a loan at a reasonable interest rate. Then net farm income dried up and interest rates rose sharply due to a variable rate clause. All his careful planning was of no avail and he had trouble servicing his loan. Murray, for the most part, did not carry land debt and was able, with some difficulty, to handle the situation. But, for others who got carried away in spending, who had large land debts, who were not skilful managers, or who just plainly got caught, the situation was devastating.

The Government of Saskatchewan passed the Farmland Security Act and the federal government established the Farm Debt Review Board. The two programs were administered jointly in Regina by Fred Switzer. These two pieces of legislation were intended to bring some semblance of order to the chaotic situation that existed. The main objective was to have a neutral party in between the lender and the borrower. This worked very well indeed; there was a loss of farms, but not as many as would have been the case without this intervention. The lenders lost money, but less than they would have otherwise.

A constant objective all during my Saskatchewan Wheat Pool involvement, and even prior to it, was to get a better return to farmers for a bushel of wheat. To this end, in 1959, a mass delegation of farmers was organized and went to Ottawa in the early winter months of 1960. We had people from every region of Saskatchewan. Most of the delegates were on the trip, but I declined to go for two reasons. First, I had a lot of livestock that needed to be cared for, and second, someone had suggested that I was just looking for another trip. The trip got attention, and the leaders were able to make the point about lack of farm income.

The Saskatchewan Wheat Pool had maintained that in the face of heavy subsidizing by the United States and Europe, Canada should bring in a system of deficiency payments. That is, a price that reflected a fair return to producers would be established and if the market did not return that amount, the government would make up the difference. This would not be an ideal solution, but a stop-gap measure to protect farmers until common sense returned to the market place. We always identified our first priority being that of a proper return from the international markets. This was prevented by the export subsidies of the United States and the European Economic Community. Therefore Canada sought international trade agreements that would eliminate or at least dilute the effect of grain export subsidies. The Saskatchewan Wheat Pool also advocated a two-price system where wheat used in Canada for human consumption would pay to producers a higher price than what could be realized internationally. This was perfectly sensible—since farmers' costs were a product of the Canadian economy, their prices should reflect that same economy. One major drawback was that, under trade rules, countries are prohibited from selling internationally at lower than the domestic price, or they can be charged with dumping. This was another reason why the Saskatchewan Wheat Pool supported an international wheat agreement.

The Saskatchewan Wheat Pool and Prairie Pools Inc. took up the issue to persuade governments of the need to restore farm net income by way of direct payments to farmers. This led to protracted discussions with the federal government. The Saskatchewan Wheat Pool was pressing for cash payments and also for the removal of subsidies through international trade agreements. The need of support for farmers by our government was intrinsically linked to creating a fair trading environment between countries that trade internationally in agriculture products, especially grains.

In 1986, there was considerable activity about the farmer income situation between farm organizations and government officials. On May 1, about twenty farm leaders met with Prime Minister Brian Mulroney in Vancouver. Mulroney was en route to Japan to attend the economic summit meeting and had stopped in Vancouver to open Expo '86.

The purpose of the meeting, at the prime minister's request, was to hear first-hand from farm leaders. He vowed to attempt to get agriculture on the summit agenda and further, that he would make our case not only in Japan, but also in several other countries he planned to visit after the summit, including China.

The prime minister was well-briefed on agricultural trade and the financial situation of farmers. He described the international situation as "a mess" because of the trade war between the United States and the European Economic Community. He recognized the importance of agriculture, a $20 billion industry, to the Canadian economy. He assured the meeting that Canadian subsidies would not be reduced unless there was a quid pro quo with other countries. He indicated that his officials were documenting all aspects of subsidization by other countries, to be used in discussion with those countries. He indicated that his government had a mission to reduce interest rates, noting that each 1 percent drop in rates would have a net benefit of $165 million to farmers.

I was called upon to make some remarks on behalf of the group, and I circulated a paper prepared for that purpose. In conclusion, I added, for emphasis, that efforts should be continued to reduce interest rates and that vigorous efforts be made to get the United States and the European Economic Community to reduce or eliminate agricultural subsidies. I also added that this was important, since we could not match their subsidies; our only weapon was persuasion. All others spoke, stating the concern of their own organizations; with the exception of the Barley Growers, all supported our paper.

John Wise, the minister of agriculture, assured the meeting that the government was not taking the situation lightly, but that in spite of its best efforts to address the shortfall of farm income, the government was being hammered financially. He warned against mentioning $2 billion as a deficiency payment because it might turn off other areas of our economy. He also reminded us that Canada did not have the ability to take on the United States Treasury.

It was a good meeting because it delivered to the prime minister and his cabinet colleagues a message of great concern. During the summer, the Saskatchewan Wheat Pool continued to work with the Ottawa office of Prairie Pools Inc. and the Canadian Federation of Agriculture to find a solution to this dreadful income situation. Our research led us to conclude that a deficiency payment of $2 billion was required.

We were invited to meet with the prime minister on September 25; essentially, it was the same group that had attended the May 1 meeting. Mr. Mulroney reported that he had been successful in getting agriculture on the summit agenda, and after a few trade-related remarks, he switched to the farm income topic and the consequences of the United States' subsidy on innocent countries, rather than on its real target—the European Economic Community. He stated that his government was aware that farmers were being hurt by actions over which they had no control. This was a significant statement because, for months—even years—we had urged our government to intervene and to have these discriminatory subsidies removed, or to compensate Canadian farmers for the economic consequences.

I pointed out that our figures indicated that a 15 percent drop in production was giving us an 18 percent lower return than we got in 1982–83 in real terms, and 37 percent drop in purchasing power. A deficiency payment would be required at a level that, when combined with the market price, would give a satisfactory return to producers.

There was an indication that cabinet ministers had accepted the need for a deficiency payment of between $1 billion and $2 billion. There was a plea by government officials to continue to work together to solve this vexing problem. This was a good meeting and encouraging in that a deficiency payment was almost a certainty.

We were once more summoned to a meeting on October 16 in Ottawa, this time to meet with John Wise, the minister of agriculture. The purpose of the meeting was to refine the distribution of a deficiency payment. The Canadian Federation of Agriculture membership was well-represented, and there were other groups from across Canada. Wise confirmed that the cabinet had accepted the need for a deficiency payment, and this was reflected in the Speech from the Throne and in remarks by the prime minister. However, he cautioned that the government was facing a deficit of $30 billion, and a payment of this magnitude might cause a run on the Canadian dollar, or a rise in interest rates.

The challenge would be to develop a formula for distribution to farmers that would have fairness and equity as its major components. As well, it would have to recognize various commodities and the uniqueness of different areas of Canada. Wise indicated that the provincial deputy ministers of agriculture had been asked to deal with the question of distribution and to make a recommendation in that regard. Wise continued that $1 billion was as far as the government could go; any higher would touch off adverse reaction by the general public. He held out the prospect that this payment could be repeated in future years until the subsidy situation was resolved. I indicated that our figures clearly showed the need for $2 billion to keep us on equal terms with American farmers. Wise responded that the amount could reach $1.5 billion if the provinces also contributed, but this was highly unlikely.

I asked for confirmation that there would be a minimum of $1 billion for the grain and oilseed sector, that it be new money, and that it would not result in a cut-back to established agriculture programs. Further, I stated, "Do not let us go out of this room believing it is $1 billion, if it is not."

There were other comments, and Wise assured us of his continuing involvement and that he would be the umpire, if necessary. Bill Hamilton, secretary of the Canadian Federation of Agriculture, asked how our committee could work with the provincial deputy ministers, but received no answer. In his concluding remarks, Wise said he was pleased with his meetings with this group and that he might recall us, even in two weeks' time, if necessary.

In December, I undertook a media tour of Ontario to make the case for agriculture. I focussed on the income shortfall for farmers, and by doing so, helped the federal government in justifying its $1 billion deficiency payment. In addition, I dealt with questions on a broad range of agricultural issues. This series of meetings was a prime example of how farm groups—in this case, under the leadership of the Canadian Federation of Agriculture—could influence positive and effective change in public policy for agriculture. The outcome of these discussions was the passing of "The Special Canada Grains Program" in 1986, to facilitate a transfer of $1 billion from the federal government to grain and oilseed producers in Canada. In 1987, an additional $1.1 billion was paid out under the same program.

The justification for the transfer was publicly announced as being to offset the losses incurred by producers resulting from the subsidy war between the United States and the European Economic Community. It was gratifying to see the government respond so quickly in a significant way to the needs of suffering grain and oilseed farmers.

CHAPTER 12
Ancillary Companies

BIORIGINAL

In 1989, following my retirement from Prairie Pools Inc., I was asked by Dave Sim, director of Plant Products Branch, Saskatchewan Department of Agriculture, to come to his office for a chat. He wanted me to review some of the projects they were planning and to offer my opinion on the viability of the program, the degree of acceptability by the public and farmers, and to offer suggestions to upgrade the plans.

I readily agreed to do so, and over the next few months, I did just that. Then Dave and I started brain-storming on how the processing of crops already produced in Saskatchewan would add value to that crop, to the benefit of producers and the province. While engaged in this exercise, it came to our attention that there was a fledgling company, PGE Canada, starting up in Saskatoon which processed borage seed. The oil extracted from the seed is classified as an essential fatty acid—apparently with properties of value to the human body—that contains Omega 6. I knew very little about such things, but knew that this borage crop and the processing of it were compatible with our value-added objective.

I was interviewed by the founders of PGE Canada, Frederick Kulow, Senior (Fred) and Frederick Kulow, Junior (Rick), a father and son who were from Massachusetts. After a fairly lengthy discussion, I learned that Fred and Rick had made arrangements with the Saskatchewan

Wheat Pool to connect them with farmers and to encourage farmers to grow borage, and also that the Saskatchewan Wheat Pool was to provide some financing. Further, the processing of the seed was being done at the Potash, Oil, and Starch Plant (POS Plant) in Innovation Place at the University of Saskatchewan campus. I agreed to join them as an advisor.

Meanwhile, Dave and I continued to assess the potential for Saskatchewan-based production. By this time, others were involved: Matt Troniak of the Economic, Diversification, and Trade Department of the province of Saskatchewan; Neil Strayer, a farmer and president of Organico, a company that he owned, which encouraged organic production and which developed markets for those products; and Bob Virgo, director of development with the Saskatchewan Wheat Pool. Also actively involved was Nuvotech, a development branch of the POS Plant. This group, with the urging of Dave Sim, agreed to formalize our efforts by organizing under the name of Vitality Health and Food Company.

I was asked to be chairman of the board and agreed to do so for a fee of $600 per day. The new company was officially incorporated as Vitality Health and Food Corporation in October 1991. We had start-up capital from Nuvotech, Organico, and the Saskatchewan Wheat Pool, but more was needed, so we sought permission from the provincial government to set up a community bond corporation. This was a method adopted by the government to encourage investment in Saskatchewan enterprises.

While playing nine holes of golf at the Wascana Country Club, I explained Vitality to Premier Grant Devine, and he was very encouraging, even to suggesting that we should seek $4 million rather than our target of $2 million through community bonds. However, before our application was approved, the community bond fund topped out at its maximum of $20 million, so it was back to "square one."

We were approached by PGE Canada to purchase Vitality Health and Food Corporation. The board of directors was reluctant to give up on its mission of developing Saskatchewan production, so the board purchased PGE Canada for $850,000 with financing coming from Nuvotech, the Saskatchewan Wheat Pool, Organico, and the Saskatchewan Development Corporation (later entitled the Crown Investment Corporation). Suddenly we were in business with Rick Kulow as the president, with product (borage oil) to sell, and an outlet for that product through Oakmont Investments, a company in the United Sates, owned by Fred Kulow.

The POS Plant was already crushing the borage seed, and PGE Canada had its office in the POS Plant in Saskatoon. The Saskatchewan Wheat Pool was a part-owner of PGE Canada by way of financial assistance, and also facilitated obtaining borage production by contracts with Saskatchewan farmers. On July 26, 1993, the board successfully merged Vitality with PGE Canada to become Vitality Health Sciences.

In 1993, we were challenged on our use of the name "Vitality" by an inoperative company in Vancouver. The board had researched the name previously and, having found no other use of it, had concluded it was available. The challenging company refused our offer of $15,000 to purchase the name. Bob Virgo of the Saskatchewan Wheat Pool and Janet Jule of the Saskatchewan Development Corporation brainstormed and came up with the name of Bioriginal Food and Science Corporation (Bioriginal).

Don Hrytzak of Nuvotech was invaluable in helping Vitality/ Bioriginal achieve its present form, and he was a very useful liaison person between Bioriginal and POS Plant. It was not long before Bioriginal outgrew the POS Plant office space, so a move was made within Innovation Place, so we were still on the University of Saskatchewan campus. POS Plant continued to provide the crushing function, but volume of product was straining the facility and timeliness of product availability was causing problems for Bioriginal. An abandoned garage at Asquith was purchased and a crushing plant for seed was set up, primarily for borage, but with the potential to crush other seed such as flax.

Bioriginal operated for several years out of the head office at Innovation Place with processing at Asquith. Then Bioriginal attained a large office/warehouse facility that had once been the Northern Drug location in Saskatoon. The crushing machinery was moved from Asquith and upgraded, which resulted in the head office, warehouse, processing, and shipping all being in one location.

Throughout this period, I was chairman of the board of directors, a very functional board representing the major shareholders in Bioriginal. I believe that my understanding of how a board should function was of great value, and the board was able to offer good advice to President Rick Kulow. In any organization, it is most important to have tight control of financial matters. Don Hrytzak had done a good job on a part-time basis, but he never had the time to devote to the full job. Rick and I were able to recruit Ron Kesseler, who proved to be absolutely superb and who gave Bioriginal many years of sound financial service.

Rick Kulow, as president, understood the science of producing Gamma Linoleic Acid, which contains essential fatty acids; this was our core business. Rick was also a superb salesman and a natural entrepreneur.

We were in the process of expanding into Europe and setting up partnerships outside of Canada. These ventures required additional capital, so we were able to bring three venture capital companies on board—Crown Life, Business Development Bank, and Manvest Incorporated. The ownership structure at this point was Crown Investment Corporation (Saskatchewan government) 17.7 percent, Manvest 11.8 percent, Crown Life 11.8 percent, and Business Development Bank 14.9 percent. In addition, Rick Kulow had 15.5 percent, while management and employees held 6.9 percent. Manvest and the Business Development Bank were Alberta-based, while Crown Life was based in Regina; each were given a seat on the board as part of their investment. The nature of the board changed and relationships needed to be built.

The Business Development Bank immediately objected to my $15,000 per year salary as chair of the board. Manvest picked up the complaint and worked to convince others that I was not an asset to Bioriginal. In the meantime, my relationship with Rick had been deteriorating because I often challenged his actions. This infuriated Rick because of his belief that the board of directors was there to offer unqualified support for management. This came to a head on two counts.

The first was my opposition to his proposal that the head office should be moved to the United States; this was supported by the Alberta-based companies but never came about. Second, I spoke against Bioriginal producing licorice to sell to a Japanese company that, in turn, would sell to a tobacco company, because it had an attractive taste in cigarettes. I opposed it on the basis that we were proudly a "health food company" and should not be supporting tobacco products. On April 17, 2000, I sent an email to Rick asking a number of detailed questions about the licorice project, dealing with the need for capital expenditures and additional costs. The preliminary reports were for a very high rate of return. Also, I expressed my concern about the impact on our image:

We portray by our words and actions that our whole function is to encourage healthy lifestyles by way of natural products, supplements, nutraceuticals, and functional foods. Will this action hurt our image by seeming to support a business that is clearly in conflict with good health?

I recorded an opposition vote when the board passed a motion to proceed with the project. This, of course, served to anger Rick and Heavanor even more.

The licorice project did not come to fruition, but because I voted against it, Rick and Tim Heavanor of Manvest were determined that I should no longer be an officer of the company. "Our other conflict, at the same time, was over relocating the head office function to the United States. When it was first presented to board members, it was easy to conclude that only the president and his staff would be transferred. A subsequent document indicated that within three years, the United States office would have the chief executive officer, the chief financial officer, the vice-president of sales, the vice-president of marketing, and the vice-president of business development—essentially the entire head office function. Left in Saskatoon would be processing, shipping, and grower contact.

I opposed this move with great vigour and was supported by Heather Forbes from Crown Investment Corporation and by Neil Strayer, also a director. This, then, provided further alienation between Rick and me. My days were numbered as a director and officer of Bioriginal. I had recently undergone a full hip replacement and lacked the energy and resolve to continue as chairman, so a short time later I resigned as chairman and as a director.

Heavanor got his goal and succeeded in being elected chairman of the board. There were a lot of funny things happening. The nomination committee was going to drop Heather Forbes from the board, even though Crown Investment Corporation was the largest single shareholder. Rick and Tim were pushing hard to bring in directors from related businesses in other parts of Canada and the United States. There would be benefits to this, if the right people were selected, but also disadvantages insofar as costs and in understanding the culture of Bioriginal and Saskatchewan.

So, after 10 years of involvement in attempting to develop and promote locally grown products, the mission was over for me but I did feel a good deal of satisfaction from helping to take an idea from inception in 1991, to fruition which resulted in $37 million in product sales in the year 2000.

The Bioriginal mission was to capitalize on Saskatchewan's natural strengths; to rapidly expand the market for natural food and health products; to diversify the Canadian agricultural crop base; to create jobs; and, to increase the value of Canadian exports via value-added processing. When the venture capital companies were added to Bioriginal, they had but one objective and that was to increase share value. They cared little about the advancement of agriculture except as it would serve their purpose of increasing share value.

The venture capital companies overwhelmed Rick, and he seemed to be fearful of doing or saying anything that was not perceived by the venture capital companies as profit-driven. The whole culture of Bioriginal changed from having multi-purpose objectives to a single objective— profit. While the last two years of my involvement were disappointing to me, the prior years had been stimulating and satisfying, and Rick and I had worked together productively, with mutual satisfaction.

I must comment on the role of Dave Sim. Dave got me involved and was responsible for encouraging the other organizations that made up the first board of directors of Vitality. Dave kept a close watch on the progress of Vitality and offered help whenever appropriate. Later, he left a "safe" job in the Saskatchewan Department of Agriculture to join the management team of Bioriginal. He served in several areas and was a major factor in Bioriginal's success. So, too, I must pay tribute to others: to Don Hrytzak, who was a tremendous help in the formative years; to Ron Kesseler, who put our financial function in good order; to Joe Vidal, who succeeded Ron and later was appointed as the CEO in place of Rick. Joe was outstanding from the very first day he joined Bioriginal.

I must give special recognition to the outstanding people who served as directors while I was chairman: Neil Strayer, who was outstanding with a natural curiosity that served the board well; Bob Virgo, who was meticulous in his approach and who had a broad understanding of the whole agriculture industry; Heather Forbes, who was always completely reliable and prepared, who brought a thorough accounting presence to the board, and was a promoter of Saskatchewan; Jim Hutch, who was a relatively short-term director who made a huge contribution, aided no doubt by extensive business and Saskatchewan experience; Rick Kulow, who brought a scientific expertise to our discussions; and Fred Kulow, who had a calming presence and who was a great source of knowledge on world markets for health food products. Of the three "new" directors, Chris Anderson of Crown Life was the only one who would have fit comfortably into the board culture that had got us to where we were.

I was proud to be part of such a fine group. We developed a culture that allowed us to be productive as we advanced Bioriginal, and at the same time to retain a compatibility that made the role of director enjoyable. The dynamics of the board changed dramatically with the addition of the three new venture capital directors. Together, these capital venture companies held 37.8 percent of the shares and thus had a dramatic influence in board decisions. Adding Rick's 15.5 percent gave the capital venture companies a majority.

In spite of all I have said, Rick Kulow deserves great credit for making Bioriginal a successful company in the health food industry and in Saskatchewan.

PLANT BIOTECHNOLOGY INSTITUTE
During my actual business life, while at the Saskatchewan Wheat Pool, or while involved with Bioriginal, I served on a number of industry-related boards. My involvement was not intensive nor was it for long periods of time.

For two or more years I was on the advisory group of the Plant Biotechnology Institute. This was a branch of the National Research Council, and unlike the Ottawa parent company, was located on the campus of the University of Saskatchewan. My input was of a general nature and not technical. There were others around the table for which the reverse was the case. I learned a lot, and contributed my knowledge about the trends in farm operations and of the needs of farmers, which may have provided some stimulus for undertakings by the Institute.

NUTRACEUTICAL NETWORK OF CANADA
My involvement with Nutraceutical Network of Canada (NCC) was as chairman of the board of Bioriginal. I could have become quite passionate about the opportunities for such a body. Nutraceuticals can best be

described as any food that has nutritious or therapeutic value and is also known as "functional food." The purpose of our committee was to persuade the government of Canada to establish a "Nutraceutical Functional Food Network." The network would be national in scope and have government support until it could be self-sustaining. The mission statement, in essence, was that the network would be defined from a regulatory standpoint and would assist in manufacturing products that would be considered legitimate by consumers. This function would, in turn, be supported by an efficient and objective regulatory environment. The goal was to have superior products recognized by the consuming public. There was the need to assist in the development and marketing of legitimate products for consumers domestically and internationally. This Nutraceutical Network of Canada was sponsored by the Western Development Fund with a commitment of $1 million and by the promise of $500,000 from the government of Manitoba.

Several conferences were held and our committee engaged the consulting firm Serecon to study the industry and to make recommendations about governance structure. The report identified the need and desire for a national network. The board submitted the report to the federal government with the proviso that acceptance of the report would entail appointing a re-constituted board. Further, we recommended there should be a review of the new board after three years' operations. Since I was representing Bioriginal on the nutraceutical board my resignation from Bioriginal ended my involvement in trying to establish what could have been a very useful entity in Canada's food system.

PHILOM BIOS

The Saskatchewan Wheat Pool was approached by John Cross, who asked for assistance in establishing a new company, Philom Bios, of which he was the president. John's intelligence and sincerity was persuasive and the Saskatchewan Wheat Pool ended up agreeing to do so. The company intended to develop products that would aid the growth and development of Saskatchewan-grown crops. This would be achieved by way of inoculating seed. The inoculate would stay with the growing plant and help it on the uptake of nutrients already in the soil, such as potash. Saskatchewan's soils are rich in potash but it is not always available to the growing plant. I was on the board for several years as the business developed and a marketing structure was established. The Saskatchewan Wheat Pool appointed Bob Virgo and later Director Rod Wiens to serve on the board of this very successful company. John Cross continually gained respect and the association with the Saskatchewan Wheat Pool was mutually beneficial.

THE CONFERENCE BOARD OF CANADA

I had the pleasure of serving on the Conference Board of Canada, a prestigious national think tank, for a few years, and was able to bring a much-needed agriculture presence to the table. The greatest return, from my point of view, was getting to know some very capable and influential people. One such individual was Ted Newall, the CEO of Dupont Canada Ltd. Ted was raised in Prince Albert and paid his way through the University of Saskatchewan by working as a cream taster at the Dairy Pool Creamery in Saskatoon. A few years later, I was able to recruit him as my co-chair on the College of Agriculture Building Capital Campaign.

ADVISORY COMMITTEE TO THE MINISTER OF INDUSTRY, TRADE AND COMMERCE FOR THE GOVERNMENT OF CANADA

The Advisory Committee to the federal Department of Industry, Trade and Commerce held very interesting meetings because the CEOs of most of Canada's leading companies were in attendance. These meetings occurred during the early years of my tenure as president of the Saskatchewan Wheat Pool. There were about 40 people in attendance, and at my first meeting I was alphabetically seated between Ian Sinclair of Canadian Pacific and Bill Twaits of Imperial Oil Ltd., two of Canada's largest companies. As individuals, Sinclair and Twaits were very influential. I was a bit intimidated, but soon discovered that they were interested in my remarks about agriculture. These were interesting meetings and helped to enlighten me about the Canadian business sector, so any benefit to the Saskatchewan Wheat Pool was not direct, but was reflected by way of my better understanding of the Canadian economy.

CHAPTER 13
The University of Saskatchewan

EARLY INFLUENCE

At a very early age I was conscious of the respected presence the University of Saskatchewan had in my home community of Maymont. The University Extension Division would come into our community and hold demonstration days on how to properly adjust farm machinery and, each year, the Mayfield-Douglas Agricultural Association would stage a banquet. A feature of the evening was a guest speaker, most often from the University of Saskatchewan Extension Division.

Lorne Paul would come quite often. He was entertaining, combining humour and a relevant message. Even though I did not understand the message in my earliest years, I knew that the university was good. My mother and grandmother always spoke highly of the university and how they wanted me to receive the benefits of a post-secondary education.

While in high school, I had set my sights on becoming a veterinarian. However, during the mid-1940s, our crops were poor and money was scarce, so plans were revised and in 1946 I enrolled in the School of Agriculture at the University of Saskatchewan. I was at university, but not at the level I had anticipated. However, it was certain by this time that I would take over the family farm, so I did not seek an off-the-farm career.

The University of Saskatchewan School of Agriculture was ideally suited for such a plan, conveying practical knowledge to a student returning to the land. Our initial class in 1946–47 had 125 members, many of whom were veterans of World War II. They brought maturity and special knowledge to our class, to the benefit of us younger participants. I had been a good student in high school and was pleased when I placed fifth in our School of Agriculture class. This two-year venture at university, besides upgrading my knowledge, gave me confidence and enhanced my interpersonal relationship skills. However, skipping public speaking classes did nothing to improve my communication ability. Then, for a period of years, my only contact with the University of Saskatchewan was to attend the "milestone years" class reunions and to pay my membership in the Saskatchewan Agriculture Graduates' Association (SAGA).

It was during my tenure as president of the Saskatchewan Wheat Pool that we began to organize day-long events involving the Saskatchewan Wheat Pool board of directors and senior management with the University College of Agriculture and the University of Saskatchewan president. These were wonderful events, beneficial to each group, enhancing the Saskatchewan Wheat Pool's knowledge of the university, and providing senior administrators of the University of Saskatchewan with farmers' insights into the real world.

I was privileged to know on a personal level many of the University of Saskatchewan presidents. They were a varied lot in so far as background and personalities were concerned, but they gave outstanding leadership through their total commitment to the institution. This is a prime example that confirms my belief that "every successful organization has leadership that is appropriate to the time." It was stimulating to work closely with Dr. Leo Kristjanson and Dr. George Ivany during their tenures as president, and with Dr. Peter MacKinnon while he was dean of law and a vice-president. As well, I was tremendously impressed with the faculty and staff members at our university.

FUNDRAISING ACTIVITIES

In 1982, President Leo Kristjanson sought a meeting with me in my office. He described his desire to build a College of Agriculture building on campus. To do so, he had to sell the idea to the board of governors of the University of Saskatchewan. To achieve a positive response, he needed to have reasonably accurate cost figures.

Leo's request was for $1,000 from me, personally, to help fund the engineering cost study. His goal was to raise $100,000 and each contributor would be called a "sod-buster." I agreed, and before leaving the Saskatchewan Wheat Pool head office, he also got a commitment from Ira Mumford, the CEO, and Jim Wright, the corporate secretary. In total, Leo was able to raise $130,000—more than enough to achieve an accurate cost estimate. Armed with the study, he was able to sell the project to the board of governors and got a commitment from the provincial government to fund the $90 million undertaking.

The province committed up to $80 million, contingent upon the University of Saskatchewan raising $12 million. In 1986, Kristjanson asked to meet with the Saskatchewan Wheat Pool board of directors to seek help for the upcoming capital campaign. He met the board in September and because his request was large, he was asked to come to the delegates' annual meeting in November 1986. His request was threefold. First, that the Saskatchewan Wheat Pool donate $1.5 million over five years; second, that the Saskatchewan Wheat Pool promote the project; and third, that I be granted permission to chair the Capital Campaign. The annual meeting approved the donation of $1.5 million over five years by resolution and agreed verbally to the other parts of Kristjanson's proposal with a resolution.

Suffice it to say that after numerous trips to various parts of Canada, after meeting with the executives of many large corporations, and after an appeal directly to farmers by a donation form in the *Western Producer*, we surpassed our $12 million target, and the best agriculture college building in Canada was completed. There were some highlights: May Beamish of Lashburn donated close to $1 million to the project. A letter from a mother and son from Marshall, Saskatchewan, contained two cheques—one for $60,000 and one for $30,000. The donation form from the *Western Producer* was the only other enclosure, and was an indication of how receptive the farm community was to support the project.

While in the head office of the Bank of Montreal, I was grilled by one of the Bank of Montreal directors about our request and how it must be easy to come to Ontario seeking help. He asked, "What are you doing in Saskatchewan to support the project?" I outlined in some detail all those businesses which had committed, and the generous response by the students of each of the School and College of Agriculture. He said, "That is good; if people don't support it where it happens, then we will not get any credit; I support your request." Also of note is that even though we concluded the campaign, money kept coming in to support the project, and by the time it was closed a few years later, over $21 million had been raised. The amount in excess of $12 million was used to add a sixth floor to the building.

Leo Kristjanson had Parkinson's disease, and some days he was scarcely able to move. I noted that on many occasions when we entered a room full of people who had granted us an audience, they seemed apprehensive as Leo came shuffling in. However, once he started to speak, they soon forgot about his affliction. His great intellect and sense of presence caused them to focus on our request. On more than one occasion, when Leo forgot to take his pills, he had to depart and leave me to finish up. He was a truly outstanding individual who, along with Premier Grant Devine, deserves most of the credit for the university having such a superb agriculture facility.

THE CHANCELLOR YEARS

In the winter of 1989, Glen McGlaughlin phoned me at my office at Prairie Pools Inc. and encouraged me to let my name stand for election as chancellor of the University of Saskatchewan. Candidates had to be a graduate of the University. The diploma in agriculture was the lowest designation to qualify for chancellor. I was elected and thus, on July 1, 1989, I assumed the role of chancellor. I had been somewhat taken aback when I told Leo I was going to contest the position; there was a long silence when I had expected enthusiasm. He later explained that he had nominated me for an honorary doctorate and wasn't sure how it would all play out.

There was no problem. I received the doctorate in May and a short time later was installed as chancellor. Three years later, I sought a second term and received it by acclamation, so from 1989 to 1995, I held an absolutely magnificent position, second only to the 17 years and seven months as president of the Saskatchewan Wheat Pool. The chancellor is also on the board of governors, so I had the opportunity to participate in all the major decisions of the University of Saskatchewan. I had the good fortune to grant degrees to many bright young people, and to see the joy on their faces as they reached an important milestone in their lives. One of my greatest pleasures was to give our daughter Jill her master's degree in horticulture in 1989. I became acquainted with the faculty and staff and marvelled at their competence and dedication. I was there when Leo Kristjanson retired and when Dr. George Ivany was appointed as the next president. Ivany and I soon established a very positive relationship that lasted through the balance of my tenure.

I became a good friend with Iain MacLean, the university secretary. Iain was an encyclopaedia on the university and he imparted his knowledge selectively and unselfishly to me. At the end of the day, we would often lift a cool one in the faculty club. Iain served the university very well indeed. Iain's secretary, Norma McBain, coached me on several fronts. As the convenor of convocation, Norma would make sure I understood my role and was equipped to do it. Everyone was very supportive of me at the university. Near the end of my second term as chancellor (an individual is only allowed two terms, then must be absent for a year prior to seeking the position yet again), the board of governors made the decision to launch a Capital Campaign to raise $30 million for a variety of upgrades on campus.

George Ivany, in concert with Ketchum Canada, was to structure the campaign. George was told by several ranking people at the university, "Don't let Turner get away." They were influenced by the success of the Agriculture College campaign a few years earlier. So, George asked me to chair the University of Saskatchewan Capital Campaign. We settled on the name "First and Best," and we were underway.

Drawing on my previous experience, I insisted that we needed to have co-chairs. In the Agriculture Campaign, I had recruited Ted Newall,

Receiving an Honorary Doctorate from the University of Saskatchewan, 1989.
Left to right: Iain MacLean, University Secretary; Ted Turner; Sylvia Fedoruk,
University Chancellor; and, Leo Kristjanson, University President.

a University of Saskatchewan graduate and CEO of Dupont Canada. Ted and I were on the Conference Board of Canada, and he delivered a good response from Ontario for us. For the "First and Best" campaign, we selected Scotty Cameron as co-chair. Scotty was a University of Saskatchewan graduate. He was prominent and well-respected in the oil patch (a term used to describe the petroleum industry) in Calgary. As it turned out, the campaign was a do-it-yourself undertaking, as even the Ketchum director Ron Strand was an alumnus of the University of Saskatchewan. The only person not an alumnus of, nor employed by, the University of Saskatchewan was Strand's assistant, Scott Smardon. Scott was younger and established great rapport with the students.

We thoroughly organized students, faculty, staff, and then the general public and businesses. Many businesses we approached had just finished their five-year pledge to the College of Agriculture. They were not impressed that we were back soliciting so soon. However, we fanned out across the country; George and I spent time in Toronto, Ottawa, Montreal, Vancouver, Calgary, and Victoria. Scotty was very successful in Calgary; there were over 500 engineers who were University of Saskatchewan graduates living in Calgary, most of them working in the petroleum industry.

In the agriculture campaign, we had focussed on the Saskatchewan Wheat Pool as lead donor, and its $1.5 million donation was about 12 percent of our goal. (As a rule of thumb, you try to get about 12 percent of the fund-raising goal from the lead donor.) After many discussions, our campaign committee felt that the Potash Corporation of Saskatchewan (PCS) would make an excellent lead donor.

George knew several of the directors of the PCS who were also University of Saskatchewan graduates. They suggested that we ask for $5 million from the PCS. George, Scotty and I, supported by other people, attended a meeting where we presented our request to Charles Childers, the president of the PCS. Childers was taken aback and said, "This is an unusually large request. How did you arrive at that figure?" There was a deep silence, so I explained the need to have our lead donor at about 12 percent of our goal, and gave reasons why the PCS was important to our campaign. The outcome was that Childers took our request to his board of directors and they approved the $5 million over five years, much to his astonishment.

Saskatchewan Wheat Pool donates to University of Saskatchewan College of Agriculture
Building project. Left to right: Ted Turner, Chancellor, University of Saskatchewan;
Garf Stevenson, President, Saskatchewan Wheat Pool; Alan McLeod, Secretary,
Saskatchewan Wheat Pool; and, Brian Whiteside, Chair, Board of Governors,
University of Saskatchewan.

The lesson here is that successful fundraising requires intense home-
work. The results of the hard work of many were successful and when
we surpassed the $30 million target, we shut down the campaign. Our
main leaders were showing signs of fatigue and many had to return full-
time to their regular university jobs. However, the fund remained open
for continuing contributions and was closed a few years later when the
amount contributed had exceeded $52 million. The agriculture cam-
paign and the First and Best campaign raised significant money for
the university, and also signalled to the public that the university was a
good place for charitable giving.

This was my last major contact with the University of Saskatchewan,
other than donating to the annual campaign. While chancellor and on
occasions since, I have helped to organize alumni events in the Palm
Springs area, which featured the University of Saskatchewan president
as guest speaker. These events have been very successful, and gave the
graduates working in the area a chance for face-to-face contact with
university people. From my involvement in fundraising for the agriculture building through my chancellor years, to the conclusion of the
First and Best Capital Campaign in 1995, I had a wonderful association
with my alma mater.

CHAPTER 14
My Community Involvement

THE GOVERNMENT OF SASKATCHEWAN, MLA SALARY REVIEW

In 1988 I was asked to serve on a committee to review the salaries of all members of the Legislative Assembly (MLAs) in Saskatchewan. The committee, deemed to be of independent citizens, was asked to review salary levels and to make recommendations for change to a three-person committee established by the Legislature, consisting of Bob Andrew (Conservative), Ralph Goodale (Liberal), and Murray Kuskie (New Democratic Party). The independent citizens' committee included His Honour Justice Ted Malone (Chair), George Solomon, and me. Gordon Barnhart, clerk of the Legislative Assembly, was the resource person to our committee.

This was a very intensive undertaking. We reviewed the salaries of every provincial administration in Canada, as well as salaries paid to the mayors and council members of some cities. We found that the most meaningful comparisons were with the other provinces. There were wide variances and Saskatchewan seemed to be at the lower end in most positions. We took into account sessional allowances and tried to make the comparisons fair. Our recommendation was that the premier's salary would be the figure from which the other salaries would be determined. We felt that the premier's salary in Saskatchewan was shamefully low.

The committee recommended that every elected MLA would receive a salary of $42,400, which was an increase from $35,211. In addition to this base of $42,400, the following positions would receive: the premier $50,000, an increase from $37,265; the deputy premier at 80 percent of the premier's salary ($40,000); cabinet ministers, $40,000; and the leader of the opposition at 70 percent ($35,000); the speaker of the house at 40 percent ($20,000); the leader of third party at 35 percent ($17,500); and, legislative secretaries at 15 percent ($7,500).

The report was well-received and adopted. While time-consuming, it was of short duration. Malone and Solomon were great to work with, as we did our short stint of public service.

THE HOSPITALS OF REGINA FOUNDATION

In May of 1997, I received a phone call from the office of the Hospitals of Regina Foundation (HRF), asking if I would consent to an interview about health care in Regina. A few days later, I went to the HRF office and met with Ron Fairchild of Navion, a fundraising firm. I knew, after the first few questions, that he was doing an assessment to determine if a capital campaign would be successful. I told him that I thought Regina was ready for such a campaign and that there was never an ideal time, so if the need was evident, to go ahead. His final question was, "Are you willing to be involved?" I responded with, "Yes, if the decision is to go ahead, I will help."

I was conscious of two facts: first, I had decided when I ceased employment on a regular basis, that I would be willing to devote the majority of my volunteer time to two areas—education and health care. I had spent time in helping to organize a National Post-Secondary Education Seminar in Saskatoon, and then served six years as chancellor of the University of Saskatchewan, so I had contributed in the education area. I had not done more than canvass for and contribute to any health-related charity. Second, my efforts in education were centred in Saskatoon. It was time I became involved in charity work in Regina.

This interview slipped from my mind until one day I received a phone call from Bob Terichow, a board member of the HRF, asking me to meet with him, Sharon Natrass, chair of the HRF, and my friend Garf Stevenson, who was chair of the Regina Health District. Once assembled, the interview came back to mind. Sharon advised that the HRF board had decided to launch a capital campaign and further, that they wanted me to be campaign chairman. I was startled. I recalled telling Fairchild that I would help with such a campaign, but I was not thinking at the level of chair; I knew how much time is involved in such an undertaking. I estimate that I spent 400 hours on the capital campaign for the College of Agriculture Building Fund, and close to 1,200 hours on the University of Saskatchewan's First and Best Campaign—and that was not counting travel time. My ultimate decision to accept the HRF request resulted in excess of 1,600 hours of volunteer time.

However, I was conscious of my commitment to myself about being involved in health care within Regina. I asked several questions about what was planned, and the readiness of the HRF for such a task. The board had engaged Navion as campaign consultants, who would put a person on site to assist with organizing such an endeavour. The HRF had set a financial target of $11.5 million on the recommendation of Navion—a figure that was determined upon from its community interviews.

The board had selected a campaign "cabinet," an industry term for "committee." I was well-pleased by the cabinet names because I knew the people either directly or by reputation. I asked for only a few days to make up my mind. There were two things that I had to get into perspective: First and foremost, when I chaired the College of Agriculture campaign, I was knowledgeable about agriculture; when I chaired the First and Best campaign, I had a good grasp about what happens at a university; but I knew very little about the functioning of Regina's health care system, nor did I know the jargon of the industry, which is so important in the area of communication. Communication is the key to a successful campaign. Second, but of lesser concern, was that I knew very little about the HRF. I knew some of the board members; I was acquainted with staff member Judy Davis, but I did not know the CEO.

My attitude toward the HRF had been influenced by Reg Johnson, a former chair of HRF and a successful businessman. I knew he was very busy, but he made time to be involved; therefore, I concluded that the HRF must be worthwhile. I agreed to deliver my answer to the campaign cabinet and to let them, after discussion with me, make the decision.

In my meeting with the cabinet, I insisted on having access to the HRF staff, on the right to organize the campaign as I saw fit, in consultation with Navion and John Dawes, CEO of the HRF. I suggested we change the campaign target to $12 million from $11.5 million. I indicated that I would make a financial commitment, and that I expected, but would not police, everyone on the cabinet to also commit financially to the project. I would not receive any remuneration for my time, but would bill expenses, with travel being compensated at the HRF rate. The cabinet readily accepted my terms and appointed me as chair; I expect they didn't know who else to approach, so on July 7, 1997, I set out on this new mission.

The first task was to become acquainted with the small staff. The offices were located on the eleventh floor of the Plains Hospital. I had indicated to the cabinet that I would develop a "case for support" and submit it to them for review and amendments. The campaign advisor, Navion, had the responsibility to draft the case for support using information gleaned from its community interviews, staff, the HRF board and Navion's own understanding of Regina's health care situation. When I asked Navion for the case for support, they handed me a 50-page document; I was expecting perhaps a maximum of 10 pages. I was startled when I read it, as Navion had taken the case for support for the Niagara Region in Ontario and had simply changed the name to "Regina" wherever necessary, except they had missed a few changes and some of the material did not fit the hospitals of Regina circumstances.

Over the weekend, I reduced the document to eight pages and felt that was plenty. The campaign committee approved it as the case for support for the "Life is Worth Giving Capital Campaign for the Hospitals of Regina Foundation." The feature pieces for purchase in our campaign included an MRI (Regina's first); two spiral CT scanners; and a new catherization lab. A case for support is important because it is adopted by the key people in the campaign structure; it means everyone is starting from the same place and working toward the same goal. It is also the main reference document in planning solicitation proposals; therefore, the utmost care is required in putting it together; and, it is essential that all of the campaign team have input into its development.

There was a most fortunate occurrence the very first day I was in the office. IMC Kalium (today's "Mosaic") contacted the HRF to see if it could make a further donation. Apparently this company, based at Belle Plaine some 20 miles from Regina, had unspent money in its charitable account. Rather than responding by phone, I set up an appointment, and then Judy Davis, one other staff member and I drove out to see IMC Kalium representatives.

CHAPTER 14

The potash company's plant manager, Norm Beug, and some of his colleagues hosted us, and we asked them for a donation of $1.5 million over a five-year period. To say the least, they were startled. Norm responded that such an amount was well beyond his ability to authorize. We explained that the HRF had just launched a capital campaign to raise $12 million and we wanted IMC Kalium to be our lead donor. I said, "I expect you have to contact John Huber, Kalium's president, in Chicago." They were surprised that I knew Huber. I had previously contacted him with success for the University of Saskatchewan's First and Best Campaign. We left on the understanding that they would relay the message, and I would write in detail about our campaign, specifically about our request. This I did, and followed it with a phone call to John, during which I offered to fly to Chicago. John said, "Don't spend money to come here. I am in Regina periodically and will let you know in lots of time, when I am ready to chat further." This was July, and in early October, his office asked that we accommodate a meeting in the Hotel Saskatchewan the third week of October, in a room large enough for 15 people, and to have PowerPoint equipment available.

This we did, and along with John Dawes, Sharon Nattrass, and Joy Dobson, we met with John Huber and he outlined what IMC Kalium was prepared to do: (1) they would donate $300,000 per year for five years; (2) they would encourage their employees to donate and would match such donations; and (3) they would encourage all of their suppliers to be generous in support of the campaign. We were overwhelmed and delighted, and thus started a relationship that lasts to this day. Norm Beug and I became good friends and enjoyed golfing together at the Wascana Country Club. Norm, a few years later, became a valued member of the HRF board and Kalium/Mosaic continued as a generous supporter. So, we were off to a good start.

Navion's performance was shaky at the outset but was satisfactory later on. The "cabinet" performed very well and I was blessed with Bob Ellard and Sue Barber as vice-chairs of the campaign; we met for breakfast one morning each week. Bob and Sue recruited Frank Hart to chair the Major Gifts Committee while Dr. Joy Dobson and Dr. Roberta McKay offered sound advice and effective contact with Regina's medical professionals. Laura Soparlo was outstanding in connecting with nurses and other hospital workers. The HRF board itself was very supportive; Chair Don Edwards was always willing to help out.

We recruited probably Regina's most respected media person, Johnny Sandison, as communications chair and Garf Stevenson offered valuable assistance as chair of the Regina Health District. Don Black hosted the cabinet meetings and gave sound advice.

The HRF staff was small but very effective, and while the campaign added to their workload, they never complained. John Dawes, CEO, was new in the position, but effective in our campaign. Judy Davis was perfectly suited to the campaign challenge and we worked effectively together as we progressed with the campaign. Due to hard work, a team approach, a compelling story, and the involvement of influential people, we surpassed our goal, raising slightly more than $13 million. There were campaign highlights such as a telethon where we were given the use of CKRM Radio from 9 a.m. to 6 p.m. one Sunday, but had to design and staff the program ourselves. With a little help from me, Judy did an outstanding job, and at the end of the day, we were $77,000 richer and those who took time to listen had a much better understanding of the Regina health care system.

Besides providing the promised equipment, the campaign raised the image of the HRF to a level that would encourage financial support and also made people want to be involved as volunteers with the foundation. Edie Hozafel, who joined the HRF staff shortly after the start of the campaign, was a manager until 2012, when she accepted the position of CEO of the Saskatchewan Cancer Agency.

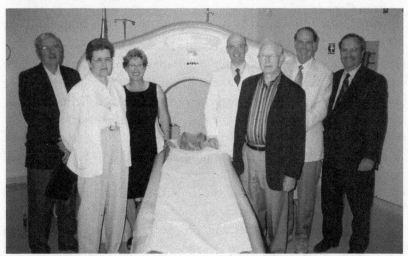

Hospitals of Regina Foundation donates a spiral CT scanner. From left to right: Dr. Ed Busse, unknown, Judy Davis, Dr. Paul Schulte, Ted Turner, Norm Bradshaw, and Bob Watt.

There was a new level of co-operation achieved by the HRF and Regina Health District. The awareness of health care needs by the government of Saskatchewan was greatly enhanced by contacts made with campaign personnel. Another most significant benefit was the feeling of accomplishment by the board of directors and the HRF staff. This new level of confidence would serve the foundation forever, and for many people, the foundation became the charity of choice. There is no doubt the Life is Worth Giving capital campaign was a spring-board that took the HRF to new and exciting levels.

In addition to those already mentioned, Laura Soparlo organized an effective campaign with staff members of Regina hospitals that resulted in strong financial and moral support. Laura was a valuable asset to our campaign. So, too, was Dr. Joy Dobson, who was most persuasive in garnering support from the physicians, as well as offering sound strategic advice. Dr. Chris Kumar, one of Canada's most respected neurosurgeons, generously supported the campaign and inspired many to contribute. Dr. Edward "Ted" Busse is one of Canada's leading cardiac surgeons. A Saskatchewan farm boy, he is completely dedicated to improving health care in Saskatchewan. He urged the HRF to upgrade the new cath lab with state of the art equipment, with the promise that if such were to be done, he would recruit bright young cardiac surgeons and support personnel. The equipment was supplied and Busse delivered, so today, Regina General Hospital has one of the best cardiac units in Canada.

Following the wrap-up of the campaign, I was given the opportunity to serve on the foundation's board of directors. This was a privilege and I enjoyed continuing in this important area of health care. Major challenges that we faced at the start of our 1997 campaign were fading away, so we no longer had to explain the closure of the Plains Hospital, nor to deal with "why doesn't the government fund these projects?" It was a great and rewarding experience. I was able to witness the growth and development of John Dawes, Judy Davis, and Edie Hozafel as well as to observe the very capable staff people who joined the foundation.

John Dawes retired from the foundation and Judy Davis was appointed as CEO. By building on the base John had established, by careful selection of personnel, and through a positive relationship with the Regina Qu'Appelle Health District, Judy has taken the foundation to success levels that are quite amazing. In this journey of achievement, she has been fully supported by an excellent board of directors, board chairs (all volunteer), and a dedicated staff.

As well, it was satisfying to see a steady upgrading of equipment in our hospitals; in some areas we are near the top in Canada. The HRF continues to attract capable people to serve on its board of directors. The year 2010 marked a cumulative $100 million raised for the benefit of Regina's hospitals—an amazing success story.

THE NATIONAL FORUM ON POST-SECONDARY EDUCATION
In 1987 I was asked by Sylvia Fedoruk, chancellor of the University of Saskatchewan, to sit on the organizing committee for a National Forum on Post-Secondary Education to be held in Saskatoon in October.

This proved to be a very interesting event. Brian Segal of Ontario was chair and Sylvia served as one of the vice-chairs. There were some 36 people on the organizing committee. It was a huge task to identify meaningful topics for discussion during the two days of conference. We wanted to have a fair representation by each province and territory for each of the 10 sessions. This proved impossible because we ended up apportioning the numbers based on provincial populations.

I tried to modify the ratio so every province would have at least one person in each session. Ontario and Quebec qualified for 40 delegates and shot down my proposal, arguing that they should not be discriminated against just because they had a larger population. In spite of the physical imbalance, however, the more than 300 attendees felt the seminar to be a huge success. I gave the closing speech to the conference entitled "The Agriculture Experience in Western Canada."

THE ROTARY CLUB OF REGINA
Since 1968, I have been a member of the Rotary Club of Regina. I joined because it seemed to be the thing to do. The Club had in excess of 140 members at the time of my joining. In some respects, that is too large to develop bonds of fellowship with most of the members. However, it was good because there were always plenty of workers for every project that was undertaken. In Rotary, I met many of the most influential people in Regina.

I stayed in Rotary because of the motto of "Service Above Self," which produced a wonderful creed when combined with four questions: (1) Is it the truth?, (2) Is it fair to all concerned?, (3) Will it build good will and better friendships?, and (4) Will it be beneficial to all concerned? Rotarians are expected to have these four questions influence everything they say and do. The Rotary way is highly ethical and the Rotary experience promotes integrity in all aspects of an individual's business and personal experience. In addition to the creed, I found members to be friendly and helpful. To this day, more than 46 years later, I still retain and cherish many of those friends. Prior to moving to Regina in 1966, I would be in Regina for a board meeting of the Saskatchewan Wheat Pool each December from 1960 onward. On more than one occasion, I was able to attend an evening of seasonal entertainment in Knox Metropolitan United Church, location of the Rotary Carol Festival. The festival, besides offering appropriate Christmas music, raised money to help needy families in Regina to have a better Christmas than would otherwise be the case. I worked at the Carol Festival virtually every year, and for the last 30 years have participated in various leadership roles. In 2010, the Rotary Carol Festival marked its 70th consecutive year.

There has never been a shortage of worthwhile Rotary projects in which to be involved, from feeding the hungry, to sending students to overseas experiences, to hosting people from other countries. One unique event was when our club supported a project initiated and promoted by the Rotary Club of Vernon, Texas. That club has marked the Great Western Trail with a concrete post strategically placed every eight miles, from Mexico, north through various states, to Val Marie and Regina, in Saskatchewan. This trail is the path that herds of cattle and horses followed when driven north to virgin lands. This took place in the latter part of the 19th century. The Rotary Club of Regina was responsible for the placing of a post and a plaque in Regina on its exhibition grounds near the Canadian Western Agribition office, and at Val Marie, near the local museum. It was a great experience, and very stimulating to see the enthusiasm of our Texas counterparts.

Rotary has helped to shape my life, has encouraged me to always be honest in business dealings, and to live in a way where I have never had to apologize. Rotary has given Mel and me a host of wonderful friends; we are grateful.

THE SASKATCHEWAN HONOURS AND AWARDS ADVISORY COUNCIL

In the mid-1980s, the Saskatchewan government decided that it would be appropriate to honour Saskatchewan people who had made a significant contribution to developing our province. To assist in this function and to protect the selection process from political interference, an honours advisory council was appointed. The lieutenant-governor was named chancellor of Saskatchewan Honours and Awards. I was privileged to serve two terms on the advisory council.

In the latter part of the 1980s, I served five years as a member. In 2001, I completed five years as chair of the council. This was very rewarding service, but did demand a significant time commitment. The citizens of our province responded with enthusiasm in nominating people to be honoured. I believe that 75 was the lowest number of nominees that I had to review in any one year. Early in July of each year, members of the council would receive a large book containing the completed nomination forms, supporting material, and letters of reference. We then had until the first week of September to become familiar with the resumé for each candidate. The advisory council would then meet, and after several hours of discussion, come up with the required number of nominees to recommend to the provincial secretary and chancellor. Each

Saskatchewan Order of Merit awards night, 2001. Front row, left to right: Ted Turner, Premier Lorne Calvert, Prince Charles, Lieutenant Governor Lynda Haverstock, Harley Olsen, and Speaker of the Legislative Assembly of Saskatchewan, Myron Kowalski.

year this seemed like a daunting task, and I was always amazed at how well the system of give-and-take between council members worked.

During my two periods of service, it was my good fortune to have Michael Jackson as chief of protocol and secretary of the council. Michael had, and probably still has, the ability to summarize the achievements of each award recipient in wonderfully descriptive language. As well, he introduced several new items to the council agenda: youth awards, a chain of office for the chancellor, and that the chancellor/lieutenant-governor should automatically be a member of the Saskatchewan Order of Merit. Michael liaised with other provinces and the federal government to ensure there was compatibility in our honour systems.

It was delightful to read the resumé of each nominee and to recognize the dedication and ability of volunteers in our society. It was also satisfying to receive credible nominations from every area of Saskatchewan.

THE SASKATCHEWAN ROUGHRIDERS

We arrived in Regina in 1966 and in 1967 we purchased season tickets for our provincial football team. We have done so each year since and the year 2013 marks 47 years as season ticket holders. What wonderful diversion to become emotionally attached to the 'Riders. It is sometimes hard to remember that it is just a game!

Mel and I have had marvellous entertainment and have stuck with the team through bad times and good times. We always said "they don't have to win to gain our approval, but they must compete hard." In recent years, our demands have been met and our expectations for success exceeded. We are so proud of the character of our coaches and players—fine mentors and delightful young men. The Roughriders are a provincial institution and serve as a rallying point for Saskatchewan patriotism.

THE WASCANA COUNTRY CLUB

I grew up with sports and was particularly fond of participating in whatever sport was available. I was introduced to golf prior to moving to Regina. Less than two years after arriving in Regina, I became a member/shareholder of the Wascana Country Club. I managed to play between 20 and 25 games per year and greatly enjoyed each outing. Mel joined in 1970 and we have continued our membership until the present time. Mel could have been very good at the game if she had enjoyed it as much as she did curling. To my amazement, I quickly advanced at the game and was fortunate to win many team events.

Following my retirement from the Saskatchewan Wheat Pool, I accepted a nomination to serve on the board of directors of the Wascana Country Club, and then in 1992 I served as its president. At the start of my term, I identified four goals for the year and successfully fulfilled them: (1) to correct the operational deficit of the previous year, (2) to successfully host the Canadian Professional Golfers' Association Tournament, (3) to rewrite the bylaws to allow women to become shareholders, and (4) to complete the renovations to the clubhouse. It was great to give back to a club that had given to me. I often described the Wascana as my "sanity pill" in days when the pressure of work was quite high. I would go out to the course and play; after a few holes, the stress would subside and I was a relaxed individual.

One of my goals as a golfer was to shoot my age. After coming close many times—for instance, when I was 78 years of age, I shot 79 several times. On May 17, 2006, I was 79 years of age and shot 77. I have, as of October 2013, shot my age or better (lower), a total of 62 times. I have had great pleasure from playing the game and from being involved in the activities of the club. I was involved in the activities planning committee as Wascana celebrated its 100th birthday in 2011. At the gala wind-up event for our centennial year, on September 16, I was inducted into Wascana Country Club Hall of Fame. I was overwhelmed by this recognition of my many years of participation and leadership as a member. I was an average golfer, and now hang on a wall with many who excelled at the game.

The Wascana Country Club has been important in our lives, providing the stimulation of playing, the pleasure of social events with friends, the opportunity to make new friends, the use of the facility for important family milestones, and a comfortable setting for business functions.

WESLEY UNITED CHURCH

Within two weeks of arriving in Regina in September 1966, our family attended Wesley United Church, which was only about six blocks from our new home on Parker Avenue. This was a natural thing to do because we had all been involved at Sharon United Church in Maymont and had attended there regularly. Why Wesley? In addition to being in close proximity to our house, we knew it was where many Wheat Pool people attended—people who would be my colleagues at work. The first impressions were good, and at the first opportunity, we had our membership transferred from Maymont to Wesley.

Mel soon became involved as a Canadian Girls in Training (CGIT) leader. Janice, Joy and Jill joined the appropriate youth groups commensurate with their age. Jack Towers, the minister, was about our age so communication was easy and we were comfortable in this new church. A few years later, when Towers left Wesley, Jill was on the search committee that hired David Iverson. Here again, we had a comfortable relationship. Mel continued her involvement by membership in the Susanna Wesley Ladies' Group until that group no longer existed. It seemed that she was treasurer of many functions at Wesley. I had been on the board of stewards at Maymont, so I accepted the opportunity to become an elder, and was on the board of trustees for about 20 years.

The congregation went through a period of instability with the ministers, but that corrected itself and we have been well-served in recent years. Wesley is an amazing congregation, filled with very capable and willing people. After the first two years, there was never a time when I was not involved in some leadership position. Mel has been involved from beginning. John Haas arrived as our new minister in 1994 and was an instant hit with the congregation; John and his wife Carolyn McBean retired in 2010. John gave outstanding service to Wesley and to the wider church. I sparked a move in 2009 to have John receive an honourary doctorate from St. Andrews College in Saskatoon, with the help of Bill Stoddard. We were successful, and Mel and I watched with pride as John was so honoured in May of 2010.

Doug Craig, served for two years as minister, his sincerity, inspiration and pleasant personality encouraged loyalty to Wesley. Kim Antosh is a lay minister, working three-quarters time at Wesley. I had the privilege to co-chair the lay ministry committee to oversee her development and for her to be proclaimed by the Saskatchewan Conference as a lay pastoral minister in the United Church of Canada. She is performing in her role beyond our fondest expectations.

Wesley is our church home and has played an important and influential part in our lives. For this we are eternally grateful.

CHAPTER 15

Reflections on the Saskatchewan Wheat Pool

t is normal, after retirement, to look back on the events of a long and varied career. I held a number of positions over the course of my working life, but in all of them, I always fought to further the interests of Saskatchewan farmers. In some cases I was more successful than in others, but I always fought what I still believe to be the good fight.

However, probably because the 26 years I was involved as a director of the Saskatchewan Wheat Pool, 18 of those as president, constituting the largest single part of my working life, and because of the profound changes that have occurred to the SWP since I left, I frequently look back and wonder, "What if?"

When I left SWP in 1987 I was royally feted and was able to maintain an informal relationship with the directors. As well, having spent 26 years as a director myself, I had nothing but the greatest respect for the board of directors and what the directors had accomplished. Then, in 1993, the board rejected my participation and indicated that they would publicly oppose my actions. This about-face occurred without warning and continues to trouble me. Although I was never given an official explanation for what happened, I believe that the problems started with the Gilson Report, which I discussed earlier, in Chapter 11.

The Gilson Report recommended that the railways would be paid the transportation revenue shortfall directly at the outset, but that payments to producers would be phased in, over a period of years, to the point where the producers would be receiving the entire subsidy directly. This flew in the face of the Pool's position of insisting the money continue to be paid to the railways.

The government was determined to change the "pay the railways scenario" to a system where the producers would be paid directly. The government no doubt felt that by doing so, it would be easier at some future time to terminate the payments, should the country's financial situation make this an attractive option.

We at the Saskatchewan Wheat Pool opposed this payment to producers because we did not want a subsidy hung around the necks of farmers. This would not play out well in international trade negotiations, even though all major countries subsidized transportation (the United States subsidized its waterways, as did Europe).

The second main reason that the Saskatchewan Wheat Pool opposed payments to producers was that the proposals suggested payments be made in a manner not related to the amount of grain produced or shipped, but on an acreage basis. This would have greatly diluted the return to grain producers. The producer was going to pay more to ship his grain, and recompense would be made to people who were not shipping grain at all. For instance, livestock producers who had cultivated land stood to gain in two ways. The higher cost of shipping would reduce the price at which grain would be sold locally—a benefit to livestock producers—and then the livestock producer would share in the subsidy at the expense of the grain shipper.

The third and perhaps most compelling reason for payment to railways was that if the government controlled the purse strings, it could make the payments conditional on performance; farmers were powerless to exercise any discipline on railways.

Only opposition to these two major issues by the Pools kept them from being put into place immediately following the report. I had, without question, been the major spokesperson in this regard; first as president of the Saskatchewan Wheat Pool and in my leadership role at the Gilson talks, and later, as executive director of Prairie Pools. We had stalled government action by being constant and compelling in our opposition. The whole question of method of payment carried into the 1990s, more than 10 years after Gilson's report.

THE PRODUCER PAYMENT PANEL

In 1993, the federal government, desperate to achieve "payments to farmers," set up a forum to help achieve its goal; the Producer Payment Panel was to be the instrument to once again examine this question. I was asked by Dennis Stephens, a government official, to be on the panel as the Saskatchewan representative. I immediately checked with the Saskatchewan Wheat Pool and the provincial government to see if they would be comfortable with my participation.

President Leroy Larsen, Vice-President Ray Howe, Vice-President Senft, and CEO Milt Fair all met and agreed that I should participate, as did Glen McGlaughlin, director of the research division. Milt and Leroy offered resources to help in my role. The provincial government also encouraged my participation, so I phoned Dennis Stephens and said I would do so.

On Monday, June 7, I received a phone call advising me that Leroy Larsen wanted to talk to me. I offered to go to the Pool office but was told that he would stop at my house on his way home that evening. He did so, but not until 11:30 p.m. It was not a pleasant meeting. Larson told me that the Saskatchewan Wheat Pool board of directors had stated that I should decline to participate on the Producer Payment Panel. It angered me that the board felt it could direct my actions, especially so after I had been encouraged to act by their most senior people. I asked to meet with the board and Larson said I could have five minutes at 9 a.m. the next morning.

My mood was tempered when I read the actual resolution:

> That the Board of Directors ask E. K. Turner to decline the opportunity to serve on the Panel established by the Federal Government to implement changes to the method of payment of the Crow Benefit, and that, in the event that Dr. Turner does serve on the Panel, the Saskatchewan Wheat Pool publicly disassociates itself from his decision and service.

The word "ask" was foremost, and much different than what I had heard the previous evening. However, the words "the Saskatchewan Wheat Pool [will] publicly disassociate itself from his decision and service" hurt me deeply, considering that I had served tirelessly and selflessly for 26 years as a director and for almost 18 years as president.

I indicated to the board that I would withdraw, since I had only agreed to serve after having been encouraged by the Pool in the first place. I stated that I had no interest in a confrontation with the Saskatchewan Wheat Pool board of directors.

I pondered the situation and sent the following fax message to Leroy Larson:

Dear Leroy,

Thank you for allowing me to meet with the Board for a few minutes at 9 a.m. June 8.

I was upset when we met at my house the previous evening because I did not understand from your remarks that the Board motion was to "ask" that I decline the opportunity to serve on the "Producer Payment Panel" and it tended to sound like a directive. When I received a copy of the motion, it was much softer than I had interpreted it to be.

Later that day I advised Dennis Stephens that I would not act on the "Panel." This was not easy for me, having been encouraged by the Provincial Government and by you to do so, and having previously advised of my willingness to participate. I have no problem with the process used but do hope, as a Pool member, that the "panel" recommendations are not justified on the basis that no one raised a concern or made a strong statement about a certain area. I recognize that many feel that to even talk to the "panel" is to condone it. Surely a strong statement of principle and detailed concerns cannot be regarded as doing anything but advancing the interest of Pool members.

There is no doubt from the budget speech of April 26, 1993, that the present Federal Government has taken the decision to change the method of payment. I quote from it, "the government is committed to the reform of Western Grain transportation policies by paying funds directly to producers through the Farm Income Protection Act and by improving the efficiency of the grain handling and transportation system.

I believe we can take little comfort if a Liberal Government is put in place, since it was a Liberal Government that started the whole method of payment situation in the first place. It is a strong possibility that representation does not ensure compliance with our position—but silence assures the lack of influence completely.

There are points that can be made that will protect the situation for grain producers in Saskatchewan. There is little doubt in my mind that if the Conservatives are returned, there will be immediate action on the M.O.P. If the Liberals are elected, the direction will be the same—but the time line extended.

Such crucial considerations by the "Panel" will include:

- the Provincial split of the Crow benefit.
- dilution.
- none for four years or?
- what commodities will be included? Only grain and/or other crops and/or forage and/or animal products?
- what area will be included in such payments—only the Prairies or all of Canada?
- the wider the area and the greater the number of commodities the heavier the hit on Sask producers.
- should farm fed grain also be included? Further dilution possible here.
- the effect on international trade—how will such subsidy extension be assessed? Will we lose access to markets because of such potentially far-reaching changes?

These are but a few of the areas that require strong attention and solid representation.

Thank you for your attention. Ted Turner

While the word "ask" softened the situation, I was distressed that the Saskatchewan Wheat Pool would disassociate itself from me, especially when I would have been advancing the Pool's position. I was disappointed and bewildered on several fronts.

First, the way the president approached our discussion, and not being granted a full discussion with the board. Second, that the board was not prepared to try to influence action in support of its policy position. Third, by my not participating, the board allowed the federal government to select someone else from Saskatchewan. The federal government recruited Bill Duke, a former president of Western Wheat Growers (Palliser). Duke was firmly opposed to the Saskatchewan Wheat Pool position and favoured payments to producers.

The board and delegates found it easier to blame me for the demise of the Crow Rate than to recount that the Pool position had been supported by 122 out of 144 delegates. Nor did it explain that our action had followed our policy and, no doubt, had achieved a better situation than otherwise would have been the case without our participation.

It should have come as no surprise a few years later that the Saskatchewan Wheat Pool declined as an effective farmer company. This was inevitable when the board of directors was not willing to stand up to defend a decision that had been made by the full policy body of the Pool some years earlier. That meeting was my last official contact with the Saskatchewan Wheat Pool and it was disappointing.

Sometime after 1993 the board, which had always been the ultimate decision making body in the Saskatchewan Wheat Pool, turned that responsibility over to the CEO. Perhaps it was when the company went public and shares traded on the Toronto Stock Exchange. Clearly, the Saskatchewan Wheat Pool as a meaningful entity for farmers was in free-fall.

Farmer/patrons of the Saskatchewan Wheat Pool, in the new ownership structure, had no financial incentive to support the Pool. It was hard to believe that a company so viable and financially sound in 1987 could be in tatters a decade later. Farmer control was what our democratic governance system was all about. Farmer control is what had inspired me and my elected colleagues over all the years.

This magnificent company, operating on behalf of farmers, is what attracted superb management people to the Saskatchewan Wheat Pool. It was this, too, that ensured us of a steady flow of competent and dedicated employees. Loyalty was a foremost factor of elected officials, management and staff.

Some say a cooperative cannot match private companies in performance. Such rubbish! The Saskatchewan Wheat Pool was clearly superior to any other company in the grain business in Canada. Credit Unions dominate the retail banking system in Saskatchewan, and Federated Cooperatives is a standout in the distribution and marketing of retail/wholesale consumer goods. Co-ops thrive, but to do so, they must have sound, dedicated board members who are willing to work hard, and who are prepared to make personal sacrifices.

Since that day, more than 20 years ago, I have had an increasing number of lingering and unanswered questions. Was the financial meltdown the result of unwise investments in Poland and Mexico? If so, what happened to the traditional conservative approach of financial management that insisted upon a potential satisfactory return on investment? Why had the board approved heavy investment in two countries with unstable economies? Why was there a reduction in the number of meetings with members? I had always found that direct comments by producers were a good guide to the direction that the Saskatchewan Wheat Pool should take.

At what point, and how, did the control of the Saskatchewan Wheat Pool shift from the board of directors to the CEO? The primary responsibility of the board, as representative of its members, was to retain control of the Saskatchewan Wheat Pool. Should it come as any surprise, then, that a CEO, once in control, would slash the costs of servicing the public policy needs of producers? When did the Saskatchewan Wheat Pool change from being a strong advocate of retaining the Canadian Wheat Board, to being indifferent on the subject, and to actually opposing retention of the Canadian Wheat Board? I am mindful and appreciative of the strong boards that I had the privilege to work with; they would never have allowed any of these things to happen. Many concerned people have asked what happened to the Saskatchewan Wheat Pool, and how did it become "just another company." My only response is that the board of directors failed the very people who put them into positions of trust and responsibility. I would say to members that, "Somehow, the board gave away control of your company and that soon made it of little value to you!"

Others have asked if I was upset by the change of name to "Viterra." My response is: "No. In fact, I am glad they changed the name because the company was no longer, nor had it been for some time, the Saskatchewan Wheat Pool that I loved and served with passion for well over thirty years.

Sadly, a prairie icon bit the dust. The biggest losers ... Saskatchewan farmers!

EPILOGUE

D espite my disappointment with the demise of the Saskatchewan Wheat Pool as I knew it, when I look back over my life, I realize how truly fortunate I have been. I was raised in a home where unconditional love prevailed; where there was discipline; where truth was prevalent and expected; where effort was encouraged; where respect for individuals, animals, and property was demanded; where hard work was natural; and where disappointment was regarded as a reason to try again.

I had, as well, the good fortune to obtain a relevant formal education to prepare me for many of life's challenges, was gifted with enough common sense to make rational decisions, and had an innate ability to relate to people in a productive manner. I chose wisely my life partner, and together we raised three capable, compassionate and concerned daughters who sorted out the essential necessities for fulfilling lives.

I was also blessed with boundless opportunities, many of which challenged my competence and thus allowed me to grow on the job. While on occasion I may not have had the required capabilities, I never short-changed the situation by lack of effort. I was fortunate to realize fairly early in life that attitude is of paramount importance and learned how to roll with the punches.

I found it interesting, in writing about my experiences, that I am more upset now about what might be regarded as injustices than I was at the time. There was always too much to do than to spend time worrying about a disruption to my routine or a failure to meet my expectations or to put balm on my hurt feelings or shattered ego.

I realized early in my tenure on the executive of the Saskatchewan Wheat Pool that a board of directors, like individuals, develops a discernible personality. In fact, a board is not truly effective until it develops that personality. As board members change, so too does the culture of the board. The challenge is to maintain a distinctive approach that is consistent with the goals of the organization. Failure to do so reduces the impact that the board has in leadership.

Throughout all of my endeavours, I was often given credit for achievements when the results were really attained by a co-ordinated team effort. My gratitude to all who helped me is sincere and beyond the capacity of this space to record their individual names.

I was committed to the Saskatchewan Wheat Pool even before assuming progressively more responsible positions in its control structure. At each new level of responsibility, my dedication increased. At the end of my career, I was totally consumed by the Saskatchewan Wheat Pool, and no doubt the Pool, in turn, was heavily influenced by me. We became almost indistinguishable, one from the other.

Work and pleasure took me beyond the boundaries of Canada. My return on every occasion triggered gratitude for the privilege of living in this country.

My family and I had a wonderful life on the farm, with superb neighbours, a productive operation, and satisfying social activities. We could have happily spent our lives there, but had we done so, it would have been to deny ourselves the many outstanding experiences that stimulated us "beyond the farm gate."

APPENDIX A

My Parting Philosophy, 1987

Address to Member Relations,
Country Elevator and Farm Service Divisions Staff
Wednesday, January 14, 1987
By E. K. Turner, President, Saskatchewan Wheat Pool

It is my pleasure to have this opportunity to address a group that is so influential in determining the course and success of Saskatchewan Wheat Pool.

Most of you not only interact with members/customers directly, but also work with or supervise others who have a first-line relationship with that important, indeed, basic group who are also the owners of Saskatchewan Wheat Pool.

I have always been impressed with the sincerity and dedication of our country managers. Indeed, one of the nicest things that was confirmed for me as I assumed executive responsibility was that the sincerity, competence, and dedication was a province-wide phenomenon—that we had managers we could be proud of by any assessment or comparison. I thank you and urge that you maintain these important positive attitudes and also foster them in others.

Yesterday, Mr. Fair presented an excellent review of the commercial history of Saskatchewan Wheat Pool. He crammed into a few pages the pertinent milestones and facts about a growing organization. It is good for us to be reminded from time to time of how we grew and developed.

Change is our constant companion in Saskatchewan Wheat Pool and so it must be, because our operating environment does not stand still. The needs of our members, the way they do business, and the expectations they have are far from being static. So change we must if we are to remain in the forefront of our industry and be ever more relevant to our members.

Some interpret change as a criticism of past actions. This simply is not correct. More likely it is an endorsement that earlier decisions provided the basis on which we can continue to build and develop, and if we do it properly we will provide the building block upon which further positive change can occur.

In my address to our annual meeting in November 1986, I spent a fair bit of time talking about the need to keep moving our horizons outward and about widening the scope of our perspective. I also spoke about developing institutions and companies that will allow us to discover and exploit continually larger markets for our agricultural products. Such an approach, if successful, will help assure that both our farms and Saskatchewan Wheat Pool retain the necessary vitality and viability.

However, while in pursuit of such important goals, we need not and indeed must not ignore the day-to-day service and information needs of our members. Mr. Fair has made it quite clear that his objective is to have more meaningful contact on a personal basis between our staff and our member/customers. He is absolutely right and I am in complete support of that concept, and so is our board of directors. It will mean changes in the way jobs are done, but it will be rewarding.

During my entire tenure with Saskatchewan Wheat Pool, grain transportation has been a matter of close attention, the only variation being the intensity of the debate and the depth of the discussions. It has been of particular importance to us since it has such far-reaching implications in both the farm policy and commercial functions of our organization. I predict that this will continue for many years yet.

We have always played a major role in such debate and we must continue to do so. It has not been an easy road. Some of the most critical comments directed at Saskatchewan Wheat Pool, and at me personally, have been on grain transportation issues, and they came from both sides of the spectrum. This will, in all likelihood, continue.

It is vitally important that we carry out effective communication with our members and staff so they can understand the issues and thus appreciate the "what," the "where," and the "why" of our positions.

We have a specific transportation issue before us now in the form of variable rates. Our position is absolutely clear and correct. However, some will accuse us on one hand of not wanting lower rates for producers, and on the other hand, of opposing system rationalization.

Both accusations are pure nonsense but will not be perceived as such unless we do an effective job of delivering our message.

We can focus our position by identifying our objectives for grain handling and transportation. I will compare those goals to that of CN and the grain companies who have applied for variable rates. I will refer to them as the CN Group.

Difference Between Our Approach and That of CN Applicants

Saskatchewan Wheat Pool	CN Group
The system should evolve	System changes should be forced
Co-operation	Dictate the changes
Consultation	Bureaucratic
Producer is of primary concern	Corporate interests are paramount

You can help producers to understand that their very right to determine the shape of the grain-collection system is at stake. The CN Group is enticing producers to pass that right to them in exchange for a questionable short-term economic benefit.

The sad fact is that producers are in such great need that they are vulnerable. But what a crime it would be if they lose the right to determine the future of the grain-gathering system, then find the $1.50 carrot is really only worth 50 cents. Or, that it is only available to the other guy. Or, that it no longer exists.

Part of the misinformation you will face is directed at the Pool. The reason is simply that the Pool has been strong and vigilant, and is seen by the variable rate supporters as the biggest obstacle they face. We must oppose such rates so that we will not break faith with our members.

One of the charges is that the Pool is against a lower freight rate for producers. That is nonsense. The Pool wants a lower freight rate for producers—for all producers, not just a few.

What the Pool does want is that the *whole system* (not only selected pieces of it) be developed on a cost-effective basis, and that producers have a voice in that consideration.

The system extends from the granary to the ship—not from the elevator to the ship. The benefits of improvements to that system must flow through to the farmer—not stop in Montreal, Winnipeg, or Minneapolis, or, for that matter, in Regina.

The complete system must be examined, and the gains must be shared equitably by all farmers. We have done much over the last two decades to help system efficiency and we are ready to do more.[1]

We need to look at some branch lines where operating costs are too high, and the long-term return too low to justify the added cost of keeping them in the system. That can be a painful process, but it is responsible. It is realistic, and above all, *it is in the best interests of producers in total.*

And let us remember that the underlying issue in all of this is really one of democracy. Farmers have the right to help decide what system serves farmers best. After all, they do pay for it. That is a point the elevator companies promoting variable rates seem to have forgotten. We have not.

At the risk of being repetitive, I am going to identify a number of concerns with the variable rate proposals.

The key point is: We are strong advocates of increased efficiency within the grain-gathering system. We believe it should be addressed by all system participants co-operating in all aspects to reduce total costs. It is imperative that the total system be analyzed right from the farm to the ship and costs minimized. That producers have a voice in shaping the future of the grain-gathering system and thus achieve a sensible evolution of the system.

1. The CN Group wishes to force the system to develop as they think best for themselves without regard for other participants. They would rather dictate the system configuration than allow it to develop. This is obvious because CN has proceeded with an application for variable rates.

2. If Canadian National is making so much money from grain that it is anxious to return a portion of it to the shippers, why do they not reduce the rate all the way across their system and thus treat all producers in an equal manner? I maintain that there are not significant savings in what CN has proposed and they could essentially arrive at the same type of train loading at many originating points across the

---

1   In a speech prepared for *Farm & Home Week* (1987), Glen McGlaughlin identified many things the industry achieved by co-operation.

prairies under today's operations and pass the savings to all producers by an overall lower rate.

3. It is not a cost-saving measure by CN but rather a major attempt to establish the principle of variable rates so that they can control the future of the grain-gathering system. I doubt that CN will be forthcoming with the actual costs associated with the system since they are very secretive about such matters. Their reluctance to reveal costs has nothing whatever to do with competition with Canadian Pacific, but rather, everything to do with hiding their costs from producers and government officials.

4. Canadian National seems ready to give up revenue and then most likely at the next cost hearing will be pleading that there must be an increase in rates because revenues are insufficient. In other words, they will be asking everyone who ships on CN to provide a higher rate for moving grain in order that they can pass revenues to those who will work with them in forcing a consolidation of the grain-gathering system. Alberta Terminals have said they will use the funds to increase their own revenues. Lorne Hehn of UGG is reported as saying UGG may use the money to upgrade other shipping points. Wouldn't it be ironic if such funds were used to upgrade facilities on lines that CN wish to abandon and thus make it more difficult for them to do so? There is little indication that the refund will be passed through to producers.

5. Producers and other shippers will need to watch very closely indeed how the car supply is handled by Canadian National. I suggest that they may attempt to operate their system in a way to make sure that cars are supplied to those who are willing to support them in their bid for a forced system.

6. Variable rates could become a reality on August 1, 1987. The bait is being dangled before producers and it will be

interesting to see how many will respond to that bait. If producers decide they want to chase after a variable rate and disregard the consequences, then we must respect the decision they demonstrate by such action. It will be interesting to see if they reject this approach since they told us to oppose it.

Rural communities have a stake in this discussion. Our quick calculation indicates 126 will be adversely affected, 27 helped in Saskatchewan.

We will vigorously oppose the principle of variable rates and we will apply for leave to appeal as provided in *The Act.*

Our goal must remain to be committed to put in place the most effective system, cost effective for producers, and volume effective to meet our grain sales requirements.

I now want to change pace once again and talk about Saskatchewan Wheat Pool.

Some of you have heard, in the last two years, what I call my "3M Speech." If time permits, it becomes a "5M Address." I have used it on many platforms in Saskatchewan. Relax. I will not deliver the whole load but simply sketch it to make a couple of points.

I have used this approach to identify the most important positive things about our organization and to talk about assets we can build on.

They are: "Members," "Manpower," and "Money."

Our most important asset is our *Members,* who are also our owners and our customers. They are found in the governance of Saskatchewan Wheat Pool and in voluntary support of our organization. They are our inspiration, our reason for being.

So, too, is *Manpower* one of our basic important assets. We have competent employees in all levels of in-scope, out-of-scope and management positions. They are dedicated to the aim and objectives as set out by the owners. They provide the expertise in designing and implementing required programs and delivering the services our members need. Their dedication is exemplary.

Money is an essential ingredient in any commercial operation. We have a sound—yes, even a "strong"—financial position in spite of a downturn in earnings the last two years. Prudent management and a pragmatic approach to financial policy by senior management and the board of directors has kept our finances as a real asset.

How successful we are is determined by how well we combine the strengths of those three basic assets. (I am now expanding into my 5M pitch.) This is called *Management*. All organizations have resources to build on. All have employees and financial assets. As well, many have a manufacturing plant, or a service facility, or natural resources they exploit. Our members/customers/owners are a resource unique to a co-op like Saskatchewan Wheat Pool and, if properly nurtured, can be invaluable. This, however, provides a special management challenge.

The fifth "M" is *Marketing*—always a vital part in determining the worth and success of an enterprise—an elusive function since its requirements are always changing—adjustment is constant, as the surrounding environment is never static. It is a mistake to try and change the environment unduly. Instead, we should design our programs and approaches to take advantage of the opportunities the environment presents us. We must market our services, our goods, our opinions to all of our publics, not the least of which are our members/owners/customers.

We are all managers in this room; therefore, we all have the task of developing a marketing strategy that best meets the needs of our target groups, whether it be as yet unknown customers overseas or those whom we know best, our members.

During my experience on the International Trade Advisory Committee, I have come to know its chairman, Walter Light. Walter is retired from Northern Telecom as its chief executive officer. He had an amazing career. In a period of 15 years, he took Northern Telecom from $600 million to over $6 billion in sales annually. He did so by well-defined marketing goals, persistence and patience. The latter two may sound like contradictions, but to illustrate a case in point: he spent time over a 10-year period trying to move his products into Japan and was finally successful to the tune of $600 million. Not bad when you can market electronic equipment into the electronic capital of the world. Not a bad return on his determination and investment of time. Much of that 10-year period was spent in learning about his customers—their needs, their desires, their environment. This required many trips to that country. Many disappointments, but eventually, success.

Obviously, he had many good, helpful things to utilize: quality, innovation, financial base, good employees, good manufacturing plants, etc., but he also used a basic theory ... one that Milt Fair has been talking about and that I mentioned earlier. Light describes this theory as "GOYA"—in polite company, it translates to "Get Off Your Anatomy." He insisted that everyone go out and meet their customers or those they supervise, and thus destroy the isolation and the lack of understanding that is a consequence of never leaving your desk.

This is the mobility, the trust Milt has been pushing. It works!

We have so much going for us—the 3Ms—and we have the ability and desire to use them effectively. I know we will do so.

You have been patient indeed, tonight. I have taken liberties with your time because it is my last chance to talk to a group so vital to our future. A group I respect so much and some with whom I have had a close association for almost 30 years. I am leaving with mixed emotions; this is the best job in Canada.

Thank you for your personal support and your dedication to Saskatchewan Wheat Pool. God bless you all.

APPENDIX B
Recognition and Awards

Canadian Wheat Board Advisory Committee
— Service Certificate..1979

Rotary Club of Regina—Paul Harris Fellowship......................................1985

Toastmasters Communicator of the Year..1985

University of Saskatchewan—College of Agriculture
— Distinguished Graduate..1986

Certificate of Recognition—Grant Devine...1987

Agricultural Institute of Canada—Honorary Membership..................1988

Co-operative Order of Merit..1988

University of Saskatchewan—Agricultural Graduates' Association
— Honorary Life Member..1988

University of Saskatchewan—Honorary Doctor of Laws.................1989

Order of Canada...1990

Agricultural Institute of Saskatchewan
— Honorary Membership...1995

Saskatchewan Agriculture Hall of Fame..1995

Saskatchewan Order of Merit..1995

Saskatchewan Centennial Medal—Lorne Calvert..................................2005

Rotary Club of Regina—Paul Harris Fellowship......................................2007

Rotary Club of Regina—Service Above Self Award..............................2008

Rotary Club of Regina—Service Above Self Award..............................2009

Wascana Country Club—Induction to Hall of Fame..........................2011

Her Majesty Queen Elizabeth II Recognition Medals
25 Years...1977
50 Years...2002
60 Years...2012